WHEN FOXES GUARD THE HEN HOUSE

WHEN FOXES GUARD THE HEN HOUSE

COMMISSIONED BY WERNHER VON BRAUN

An Inside Look at the Risk of Nuclear
Annihilation in the Weaponization of Space

APOLLO 14 ASTRONAUT & ASTROPHYSICIST
DR. EDGAR MITCHELL

COMPILED BY CAROL MERSCH

First Edition
Printed and bound in the U.S.A.
ISBN: 978-1-68313-269-1

Pen-L Publishing
Fayetteville, Arkansas
www.Pen-L.com

"Human excesses and error have historically sorted themselves out in the fullness of the centuries. But never before has the capacity existed for humans to terminate all living activity on the planet in one spasmodic hour of nuclear exchange."

– Apollo 14 Astronaut Dr. Edgar Mitchell
May 10, 1987

OTHER BOOKS BY CAROL MERSCH

———

The Apostles of Apollo

The Space Less Traveled

The Incredible Reverend Stout

We Are One

We set sail on this new sea because there is new knowledge to be gained, and new rights to be won, and they must be won and used for the progress of all people. For space science, like nuclear science and all technology, has no conscience of its own.

Whether it will become a force for good or ill depends on man, and only if the United States occupies a position of pre-eminence can we help decide whether this new ocean will be a sea of peace or a new terrifying theater of war.

I do not say that we should or will go unprotected against the hostile misuse of space any more than we go unprotected against the hostile use of land or sea, but I do say that space can be explored and mastered without feeding the fires of war, without repeating the mistakes that man has made in extending his writ around this globe of ours.

– Apollo 14 Astronaut Dr. Edgar Mitchell, 1987

CONTENTS

THE PREMISE

When foxes guard the henhouse, someone must watch the foxes, else only they are satisfied.
                                                    – Dr. Edgar Mitchell, 1987

In 1974, Wernher von Braun was dying of cancer.

In both reality and myth, von Braun's career embodied the clash between World War II German Nazis and the all-American dream, the "good versus evil," and the necessary grand battle that forged the identity of the United States.

He was the Cold War personified.

Some adored him as an American hero of space. Others could not shake the Swastika invisibly stamped on his psyche.

He had learned a thing or two in his legendary life, or more accurately, a thousand things or more, and the sum total of it all was a calculus only he could determine.

All of this, in the end, made him the American icon that he was—and became the synthesis of his final request on Earth.

When he summoned Carol Rosin to his office at Fairchild Industries in early 1974, he did not mince words. Her recollection was vivid.

"The very first day that I met Von Braun, he had tubes draining out of his side. He was tapping on the desk telling me, 'You will come to Fairchild.' He said, 'You will come to Fairchild, and you will be responsible for keeping weapons out of space.' He said it with intenseness in his eyes and added that space-based weapons were a dangerous, destabilizing, too costly, unnecessary, untestable, unworkable idea."

– Carol S. Rosin, 2000

He asked Rosin to act as his spokesperson and to appear on occasions when he was too ill to speak. According to Rosin, he also asked her to take on the challenge of promoting the ban of space weapons by educating decision-makers and the grassroots population about transforming the military industrial complex into a peaceful space exploration industry.

She accepted von Braun's challenge, and in 1974, Rosin, a former schoolteacher, joined America's foremost space scientist at Fairchild Industries as its first female corporate manager.

Rosin's meteoric rise from a schoolteacher to an executive at Fairchild Industries was ignited when she transformed her Alexandria, VA, sixth-grade classroom into "Spaceship Earth." This caught the attention of Fairchild executives, and in 1974, at the urging of von Braun, she joined Fairchild and became a vocal advocate of the open and collaborative development of space technology that would further mankind rather than destroy it.

Rosin echoed von Braun's concern that an orbiting surveillance U.S. space station under the guise of a peaceful defense mechanism offered an all-too-convenient platform with which to launch a preemptive missile attack on the Soviet Union. Once initiated, the prospect for life on the planet would become dismal. Rosin made known her concerns to Washington officials and to anyone who would listen.

The potential ramification of such a proposed initiative was of extreme concern to von Braun, who had a history of working with weapons systems.

As a young man, von Braun worked in Nazi Germany's rocket development program, rising from a World War II Nazi war camp to mastermind the deadly V-2 missiles used to lob warheads onto London. Even then, he dreamed of future manned space journeys.

As World War II was ending, von Braun surrendered himself and his cadre of over 126 German rocket scientists to the 44th U.S. Infantry Division in the Bavarian Alps, May 2, 1945. Following the war, he and 1600 top scientists were secretly moved from Nazi Germany to the United States under the U.S. Army's counter intelligence agency as part of Operation Paperclip, where he worked for the United States Army on an intercontinental ballistic missile (ICBM) program and in 1958 developed the rockets that launched the United States' first space satellite, Explorer 1, into orbit.

With his unparalleled expertise in modern rocketry, in 1960 von Braun's career trajectory landed him a position as NASA's director of the newly formed Marshall Space Flight Center in Huntsville, AL and chief architect of the Saturn V super heavy-lift launch vehicle that propelled Apollo spacecrafts to the Moon.

During this time, the uneasy wartime alliance between the United States and the Soviet Union began to unravel in what some predicted would be "a nuclear stalemate between monstrous super-states, each possessed of a weapon by which millions of people can be wiped out in a few seconds." The rivalry signaled the beginning of the Cold War between the United States and the Soviet Union, and Moscow was flexing its political muscle by ostensibly outstripping the U.S. in many areas, including space technology.

The initial centerpiece of Wernher von Braun's plan in the 1950s and 1960s was an armed space station that would serve as a reconnaissance platform and orbiting battle station for achieving "space superiority" over the Soviets. One of its roles could be, if necessary, the launching of nuclear missiles.

The initiative for the weaponization of space soon faded, however, as the two superpowers decided that it was not in their mutual interest to put bombs in orbit or to threaten each other's military reconnaissance, navigation, and communication satellites. As a result, the nuclear-armed space station was quickly forgotten. And when von Braun joined NASA in 1960, it was positively at odds with the agency's mission of peaceful space exploration.

His work at NASA from 1960 to 1972, and his friendships with Walt Disney and President John F. Kennedy, among others, further turned him against placing weapons on a U.S. space station.

In 1970, NASA leadership asked von Braun to move from his home in Huntsville to Washington, DC, to head up the strategic planning effort for the space agency.

While in Washington, the rumblings coming out of the Pentagon were disturbing to von Braun. Of special concern was the potential use of thermonuclear weapons with the resultant escalation of destructive power to levels suicidal for the human race, coupled with a series of disturbing and irrational proposals for launching nuclear weapons from orbiting military stations with little or no regard for the potential political and defense risks. Such ill-conceived strategies could easily be defeated, or worse, instigate a preemptive irrational nuclear response from an antagonistic nation.

Sensing that his ideals no longer coincided with those of Washington officials and suffering health problems, Von Braun resigned in 1972 and left Washington to join Fairchild Industries, an aerospace company in Germantown, Maryland, as vice president, where, in 1974, he recruited Carol Rosin.

In the years that followed, Rosin became his protégé and confidant.

> "Over and over during the four years that I knew him and was giving speeches for him, he said there are many enemies against whom we were going to build this space-based weapon system; the first of whom was the Russians … Then there would be terrorists; then there would be third-world countries, rogue nations … nations. And then he would repeat to me over and over, the last card, the last card, will be the extraterrestrial enemy card. The intensity with which he said that made me realize that he knew something that he was too afraid to talk about."
>
> – Carol S. Rosin

During his years at NASA, von Braun developed a close relationship with Apollo 14 astronaut Edgar Mitchell, a brilliant astrophysicist with a PhD in Aeronautics and Astronautics from MIT. Mitchell was the sixth man to walk on the Moon, and the view of Earth from space gave him a new perspective on the fragility of the planet and its inhabitants. He and von Braun came to share the pursuit of peaceful space exploration—not one of nuclear confrontation. Mitchell also shared von Braun's views on extraterrestrials. He grew up near Roswell, NM, and his later visits with elderly residents who witnessed the 1947 UFO crash in Roswell convinced him the accounts of extraterrestrials were genuine.

Von Braun died of cancer in 1977.

One of his last requests to Rosin before his death was that she locate Edgar Mitchell and enlist his help in elevating public awareness of the hazards of developing orbiting military platforms.

After retiring from NASA in 1973, Mitchell became involved in scientific studies that brought him in contact with members of the Nixon, Ford, and Reagan administrations, including CIA Director George Bush, Sr. During this time, concerted efforts were underway by the Reagan administration to ascertain the Soviet's armament readiness for purposes of possible U.S. retaliatory action. The main goal of the U.S. foreign policy during the Reagan administration was winning the Cold War.

To this end, Reagan proposed a Strategic Defense Initiative (SDI), a U.S. strategic defensive system against potential nuclear attacks—originally conceived from the Soviet Union. The SDI was first proposed by President Reagan in a nationwide television address on March 23, 1983. Because part of the defensive system that Reagan advocated would be based in space, the proposed system was dubbed "Star Wars," after the space weaponry of a popular motion picture of the same name.

As von Braun had warned, Reagan's SDI initiative brought the stark issue of weaponization of space back on the agenda.

The SDI was intended to defend the United States from attack from Soviet ICBMs by intercepting the missiles at various phases of their flight and directing space-based laser killing beams toward moving Soviet targets. One of Reagan's advisors was a controversial Hungarian nuclear physicist, Edward Teller. Teller, known colloquially as "the father of the hydrogen bomb," was a vocal advocate for nuclear energy development, a strong nuclear arsenal, and a vigorous nuclear testing program. He was especially known for pushing controversial technological solutions in military operations.

It became clear to Mitchell that such an extreme posture for implementing a defensive strategy using the combination of orbital and nuclear technology could, with only minimal provocation, result in a quick and catastrophic annihilation of not only our enemy, but ourselves.

The potential for nuclear disaster was immense.

Rosin saw Reagan's SDI proposal as the first hint that von Braun's nightmare was beginning to unfold. In 1984, per von Braun's request, she approached Edgar Mitchell to assist in developing an in-depth look at Reagan's proposed initiative that could include the use of weapons in space and, more concerning, the misuse of an orbiting surveillance station armed with nuclear weapons.

With Rosin's access and research into the political underpinnings of the program, Mitchell developed a thorough and masterfully written thesis exposing the weakness and vulnerability of Reagan's "Star Wars" program—or any space-based platform armed with nuclear weapons. The manuscript was aptly titled *When Foxes Guard the Hen House*.

> Reagan's 'Star Wars' initiative to put orbiting weapons into space aroused the ire of many scientists, who realized, as did I, that weapons in space put a hair trigger on war, and more likely, a nuclear exchange. What nation would allow an offensive weapon orbiting over their territory that could be de-orbited in an offensive attack without prior notice? Certainly not any nation where there were hostile feelings, such as those between the U.S. and the Soviet Union in the Cold War. Such an orbiting device would virtually invite instant attack. Many scientists wrote letters and signed proclamations, but Reagan persisted. Carol Rosin worked as an assistant to Wernher von Braun, and before he died, he sent her to me to work on these issues. We did the research, and I wrote the first draft. Thirty-four publishers refused to publish it or even seriously consider it. Such was the power of the White House, which did not want it published. I was subject to harassment, tapped phones, and more, much like we experience with government Special Access Programs today.
>
> – Dr. Edgar Mitchell

The finished manuscript was subsequently placed in storage in Denver, Colorado, where it remained until 2003 when Mitchell retrieved it at my request. In 2006, he gave me written permission to publish the manuscript.

Conversion and cleanup of the lengthy hand-typed manuscript into a computerized document was cumbersome. The resulting unedited manuscript lay in my files for two decades as Mitchell and I pursued other endeavors.

At the request of Mitchell before his death in 2016, the manuscript was resurrected and edited prior to publication.

Edgar Mitchell died on February 4, 2016.

While much has been documented by historians and military aficionados about von Braun's earlier push to weaponize space during the 1950s and 1960s, little has been written about his reversal of intentions in the 1970s as conveyed in the first-hand accounts of Carol Rosin and Edgar Mitchell.

The sum total of his accomplishments makes von Braun the most influential rocket engineer and spaceflight advocate of the twentieth century. Five hundred years from now, humans may remember little of the twentieth century except for the nuclear bomb, industrialized mass murder, the discovery of global warming, the emergence of computer networks, the achievement of powered flight, and the first steps into space.

Ultimately, concerns about the objectives of SDI and fiscal budgetary constraints forced the Reagan administration to announce the program would cease, and it was formally scrapped in 1993, to be replaced by other military prerogatives milled in the fears of nuclear threats—real or imagined.

Assuming that we do not ruin the Earth through warfare or environmental impact, actually leaving the cradle of all terrestrial life to establish a foothold in space may, in evolutionary terms, rank among the most important endeavors yet to be accomplished. In those terms, Wernher von Braun deserves to be remembered as one of the most influential rocket engineers and spaceflight advocates of the twentieth century.

The legacy of his concerns relayed to Carol Rosin and diligently recorded by Edgar Mitchell deserves our collective attention.

– Carol Mersch, Compiler

THE LOOMING CHALLENGE

There will never be a nuclear war since we have gone 42 years without one. The world has seen what the two atomic bombs did to land targets and also observed the results of H-bomb tests. That alone might keep anyone from making the mistake of using one.[1]

— Major Thomas Ferebee, Inola Gay Bombardier
Atomic Bomb, Hiroshima, Japan, Aug. 6, 1945 (WWII)

On two days in August 1945, U.S. planes dropped two atomic bombs—one on Hiroshima, one on Nagasaki, the only times nuclear weapons have been used in an attack.

On August 6, 1945, Major Thomas Ferebee, bombardier on board the B-29 Superfortress bomber *Enola Gay*, unleashed the first atomic bomb on Hiroshima, Japan, that would end World War II. Three days later, a second atomic bomb was dropped on Nagasaki, Japan—"just to let them know the Americans have more than one," said Lt. General Leslie Groves, chief of the top-secret Manhattan project.[2]

Estimates are that nearly 200,000 civilians and over 20,000 soldiers were killed. "Practically all living things, human and animal, were literally seared to death," an Associated Press story reported.[3] Those who were

---

1 Letter from Thomas Ferebee, Col USAF retired, to Anthony Pizzitola, Houston, TX, July 21, 1987.

2 Evan Thomas, *The Road to Surrender*, (Random House, 2023).

3 "The U.S. Was There: U.S. dropped atomic bomb over Japan 1945," Associated

close to the epicenter of the explosion were simply vaporized by the intensity of the heat.

A few days later, Japan announced its unconditional surrender. World War II was over.

Atomic Bomb creator Robert Oppenheimer said he believed "it will make war so horrific that humankind will ban it."[4]

Still, those in charge knew that if the new, dark science of the atomic bomb were not controlled, civilization itself would be at risk. The United States had a monopoly then, but it might not last, and a less scrupulous nation would build its own larger, more deadly weapon.

The technology of space weaponry has grown at warp speed from the "simplistic" nuclear bomb lobbed on Hiroshima from a B-29 bomber in 1945 to escalating incidents of sophisticated space weaponry and ultrasonic warfare in the current age. "The space battlefield is not science fiction and anti-satellite weapons are going to be a reality in future armed conflicts," said Lt. Gen Shaw, deputy commander of the U.S. Space Command.

"A ship in the Pacific Ocean carrying a high-power laser takes aim at a U.S. spy satellite, blinding the satellite's sensor and denying the United States critical eyes in the sky." This stark hypothetical scenario laid out to military officials at the 2021 Space Symposium caused the U.S. Space Command to consider instigating a "space warfighting doctrine." Such an attack could lead to escalation and wider conflict as rival nations like China and Russia step up development and deployment of anti-satellite (ASAT) weapons.[5]

Military space assets like satellites and ground systems have typically been considered "support" equipment that provide valuable services such as communications, navigation data, and early warning of missile launches. But as the Pentagon has grown increasingly dependent on space, satellites are becoming strategic assets and coveted targets for adversaries.

Of particular concern to the Pentagon are disruptions to satellite

Press, Aug. 5, 2015.
4   Ibid.
5   Sandra Erwin, "U.S. generals planning for a space war they see as all but inevitable," *Space News*, Sept 17, 2021.

communications networks that are used to operate unmanned surveillance aircraft. Drones rely on GPS and satellite communications systems to track and strike targets.

China and Russia, for example, have direct-ascent weapons that are launched on a suborbital trajectory to strike a satellite in orbit. They also have co-orbital weapons that are placed into orbit and then later maneuvered toward their intended target.

According to the Center for Strategic and International Studies, China and Russia are deploying non kinetic space weapons. These include lasers that can be used to temporarily dazzle or permanently blind sensors on satellites and jamming devices that interfere with the communications to or from satellites by generating noise in the same radio frequencies. It would also evoke a material disruption to society, as satellites play a growing role in cellphones, navigation, supply chain, logistics, and healthcare.

A key reason the space race is accelerating is that technology is advancing rapidly. A second reason is the absence of binding commitments to the operating norms in space. The past decades have brought little deterrence to a nuclear conflict between global powers as a variety of national, international, and bilateral proposals have failed to receive ratification by one or more participating countries, including the U.S.

The 1967 Outer Space Treaty banned the stationing of weapons of mass destruction in outer space, prohibited military activities on celestial bodies, and detailed legally binding rules governing the peaceful exploration and use of space. The treaty was spurred by the development of ICBMs in the 1950s, which could reach targets throughout space. The Soviet Union's launch of Sputnik, the first artificial satellite, in October 1957, followed by a subsequent arms race with the United States, hastened proposals to prohibit the use of outer space for military purposes. However, nothing in the Outer Space Treaty prevents military forces from being in space, the construction of military outposts in space, or even conducting war operations in space.

The Outer Space Treaty was entered into force October 10, 1967, and 114 countries have ratified the treaty, including the U.S. and Russia, with another 22 countries that have signed it but have not yet completed ratification. Recent U.S. government-imposed initiatives have attempted to close the gap.

In 2019, space-based interceptor development resumed for the first time in 25 years with President Trump's signing of the National Defense Authorization Act, which authorized the U.S. Space Force to be an independent military branch under the U.S. Space Command. Trump made it clear that from his perspective that space will be "the next war-fighting domain." Trump's CIA Director, Mike Pompeo, called for additional funding to achieve a full-fledged "Strategic Defense Initiative for our time, the SDI II."

Experts point out that there are increasingly more ways to permanently or temporarily damage satellites so it would be virtually impossible for the U.S. Department of Defense to defend against a multitude of weapons. Without a credible deterrence capability, adversaries may be willing to gamble on relatively minimal blowback to attack and permanently take out these essential U.S. space-based capabilities.

"We've seen very intentional interference within regional conflicts to take military systems offline," said Frank Backes, senior vice president of space and defense contractor Kratos. "The only foundation of international space law that currently exists, the 1967 Outer Space Treaty, is outdated and doesn't address most space security issues that could set off a war."

The Treaty on the Prohibition of Nuclear Weapons was ratified in January 2021 and was the first legally binding international agreement to comprehensively prohibit nuclear weapons. The treaty passed with 122 nations in favor; however 69 nations did not vote, among them all of the nuclear weapon states and virtually all NATO members, including the U.S. The inaction leaves security risks for the U.S. and its major adversaries with the capability for weaponization of space—including Russia, China, North Korea, and Iran.

In 2022, President Joe Biden's administration announced it would ban destructive anti-satellite testing in space to prevent space debris and set international norms. The statement was issued as a response to the Russian direct-ascent anti-satellite test in November 2022 that destroyed one of its own satellites and created thousands of pieces of debris. The debris will be a danger to satellites and other space objects for years. The ban, however, is not believed to halt U.S. development of space-based weapons.

As of 2023, development of U.S. military's space weapons designed to deter China from firing the first shot against a satellite remain classified.

In a rare disclosure, the Space Force last year said it deployed an advanced ground-based communications jammer that could be used as an "offensive weapon" to disrupt enemies' satellite transmissions.

In late October 2023, the Pentagon announced—to the surprise of many, including congressional staffers who work on these issues—that it was pursuing a new nuclear weapon to be known as the B61-13, a gravity bomb.[7]

This was a troubling development for many reasons. First, it was merely the latest in a long line of new nuclear weapons that the United States is building or proposing, in yet another sign that a new nuclear arms race is expanding.

The looming challenge resurfaced in national headlines in early 2024 when the White House confirmed that "Russia has obtained a 'troubling' emerging anti-satellite weapon but said it cannot directly cause 'physical destruction' on Earth." Physical destruction aside, the damage incurred by disruption of satellite communications would invoke equally troubling consequences and inevitably entail the use of nuclear weapons in space.

Former CIA Director Leon Panetta noted the news was a wake-up call that Russia still remains very much a threat to the United States. "The problem is that we use satellites for communications purposes, we use satellites for intelligence purposes, we use satellites for GPS purposes and other information purposes as well that relate to our national security. If Russia could blind our ability to be able to gather that kind of information, make no mistake about it, that would be an act of war because it would threaten our national security."[6]

By April 2024, the United States and Russia were set to face off over nuclear weapons in space at the United National Security Council over a draft resolution spearheaded by the U.S. and Japan, aiming to prevent the placement of nuclear weapons in space and not take steps to develop weapons of mass destruction that are deployable above Earth. The draft resolution came after the U.S. accused Moscow of developing an anti-satellite nuclear weapon to put in space, an allegation Russia flatly denied.

Russia vetoed the draft resolution, a move that prompted the U.S. Ambassador to the UN, Linda Thomas-Greenfield, to question if Moscow were hiding something.: "Why if you are following the rules would you not support a resolution that reaffirms them? What could you possibly be hiding?"

Reflecting on the belief by *Inola Gay* Bombardier Major Thomas Ferebee that results of the two atomic bombs and H-bomb tests "might keep anyone from making the mistake of using one," Hungarian physicist Edward Teller, inventor of the H-bomb and a promoter of Reagan's "Star Wars" program, had stern words for a future where nuclear weapons remain in the arsenal:

> "I do not feel that there is any chance to outlaw any one weapon. If we have a slim chance of survival, it lies in the possibility to get rid of wars. The more decisive a weapon is the more surely it will be used in any real conflict and no agreements will help. Our only hope is in getting the facts of our [H-bomb] results before the people. This might help to convince everybody that the next war would be fatal."

Barring an explicit international treaty that bans all weapons from space, not just the "weapons of mass destruction" outlawed by the Outer Space Treaty, von Braun"s concept of an orbiting battle station to achieve "space superiority" over the Soviets will remain an issue of great importance.

– *Carol Mersch, Compiler*

INTRODUCTION

There is no strife, no prejudice, no national conflict in outer space as yet. Its hazards are hostile to us all. Its conquest deserves the best of all mankind, and its opportunity for peaceful cooperation may never come again.

— Astronaut Edgar Mitchell, Apollo 14

The twentieth century has brought changes in the human way of life greater than any previous century in history. The rate of change is itself a phenomenon which we humans have difficulty accommodating. We seem to desire to find a status quo and maintain it; to find that quiet spot where tranquility prevails in order to construct a life full of meaning. Wrenching changes in life patterns bring on stress and discomfort, and we humans resist making those changes.

Then there are some who, while not meaning to create change, want to explore things, create things, and know about things. And when certain humans engage in those activities, life for all suddenly begins to change. Our exploring, thinking, and doing result in the status quo moving to a new level which the next generation tries to accommodate and maintain as though it had existed and should exist forever. This almost schizophrenic internal conflict to maintain a stable pattern for daily activities of life but to also explore how to make life better has brought us, after 5200 years of recorded activity, to the point where all is changed, all is process. The status quo is measured in seconds, not generations as it was before this century.

Perhaps the greatest change ever, but one that has yet to be fully understood and accommodated, is the activity that has taken humans into the space environment. We have broken the bonds of gravitation and left the nurturing breast of Earth to explore the void beyond this tiny repository of life. There can be no more hostile environment for living creatures than the emptiness of space in which no single element is present to sustain life except raw solar energy. It epitomizes isolation and loneliness. Yet in that environment, in our primitive spacecrafts, a few humans have found a new vision.

The preciousness of the Earth and its life-giving properties, alone and dwarfed by the vastness of space, reaches out to inspire. Of the handful of explorers to experience the space environment and to see Earth from afar, a few have had the extraordinary experience of gaining new insight about our planet, our lives, and about the process of life itself. That insight has been no respecter of culture, creed, or politics. It has come to astronauts and cosmonauts alike and from all the 19 nationalities that have had people in space. It arises when new images and information floods an eager and open mind to reshape traditional thinking and previously held beliefs into a different framework for thought. Horizons become expanded and new possibilities for old dilemmas present themselves.

There is a risk to those who speak of such new things too boldly. The status quo delights in the ridicule of each new idea that challenges the "established knowledge" of the moment. But new insight cannot arise from rigid and preprogrammed beliefs, from those clinging tenaciously to a status quo present or previous, regardless of the stimulation from new information. And insight can be rejected or ignored if it poses significant threat to pre-existing belief. But to those willing to see, hear, and to re-think the meaning of existence, there is a message: The status quo in human affairs is only a step in the inexorable movement of evolution. Change is relentless, however imperceptibly it seems to move. And traditional human viewpoints existing for centuries, revered and hallowed, lead us into dangerous waters as we look to the past for answers that need to come from the present.

For me, most of the old pat answers steeped in rich traditions of the past do not work. At the root of our thinking are fundamental errors about the nature of existence. The two traditions of human thought, one older than the other, religious belief and scientific inquiry, both fall short of providing the necessary answers. And they have not, though needed, resolved the chasm between them, for the Universe does not give multiple answers to the same question about fundamental reality: of how chance and choice contend together to determine the future of existence.

It is with such thought that this spacefarer looked at the cosmos and the Earth from afar, seeking the oft-sought riddle of personal existence, the turning of orbits rolling us through the heavens toward predetermined points with the mind spanning the gap toward the galaxies to seek a hidden truth. And it was there, but different from that sought or expected! Not obviously and blatantly, hanging like fruit from the tree in the Garden. But subtly and mysteriously pulsing within to expand awareness of life, creativity, intelligence in an otherwise mechanical expanse. And it was not the answer one would find within tradition, if indeed one expected an answer at all, but an inexpressible awareness! That awareness flowing through Being had no expression in the traditional metaphors of the past and finds none today sufficient to the task of expressing the truth. The awareness remains and the truth remains, awaiting the words and phrases that must yet be constructed to couch an uncommon experience in common language. Like the dumb, with knowledge of words but unable to speak.

The language, forms, beliefs of the past do not suffice, but here and there an idea can be plucked from the old fabric of history, philosophy, spiritual tradition, and science to weave a new fabric of knowledge that looks different from the old. A fabric many years in the weaving, full but not complete; for the Universe is not complete as it moves through space and time. The true expressions as they become known, however, will not fall pleasantly on the ears of many, and the fabric will not clothe in warmth those who seek refuge in abdicating their personal responsibility for knowledge, choice, and change—for those venerating the myths of the past.

The past will not suffice to govern the future for the past was wrapped in ignorance and false beliefs about what is, however well those beliefs served their times, at the time. Even the most enlightened of our forebears could be no more enlightened than to express the knowledge that had evolved to their era, even when more inexpressible then than now, because the information, language, and thought were less evolved. Truth is ineluctable even when inexpressible in the metaphors and knowledge of the day. Thus, to cling to the "truth" as expressed by tradition invites error when the status quo has changed and new knowledge is abroad.

There has been for many of the space travelers a common warp and weft to that space-inspired insight. It runs through the fabric and is expressible. It is a vision of cooperation to preserve this Earth, its species, and its biosystem from further human damage. It is a vision filled with urgency and a sense of personal responsibility to arouse our neighbors to the need for reassessing our condition. Our planet is fragile and magnificent and is in great peril from thoughtless human activity but need not be. The expression of this new cooperative vision takes many forms, as many as there are people to express it. And that space inspired vision is no longer confined to the handful of space explorers who have physically seen the Earth from space, but has magically touched many others in diverse and wonderful ways.

The time is now upon us where the humans on planet Earth must consciously choose our destiny. The consequences of our large-scale actions must be examined. Do we attempt to maintain the status quo or, worse, attempt a return to the thinking and norms of yesteryear, as some would wish? Or do we recognize that our activity of discovering the knowledge and creating the technology in the twentieth century have left no alternative but to accept new responsibility for the consequences of our evolving presence in the biosphere?

The metaphor of the title: "When Foxes Guard the Hen House" can be taken in many ways. Foxes are imbued by folklore with cunning and stealth. Yet they operate mostly by instinct, not reason and insight. Were there only one clutch of fowl left on Earth, the fox, to his own detriment, as well as that of the fowl, would devour them. Likewise we humans, when we blindly follow our instincts and traditions, rooted in the lessons of history, behave no better than the fox.

One of the pervasive traditions of human behavior is the veneration of conquest and war as a necessity of existence. However traditional and instinctive it may be, however much conventional wisdom resists such a fundamental change, it is a human activity that must be outgrown and placed behind if the species of Earth, all species, are to survive. But it is not a behavior that will be changed quickly or easily. Evolution has never worked its wonders rapidly. But change it must for there is no other way to exist as the populations crowd the Earth and destructive technologies increase their power. Many species of this planet have disappeared as they failed to adapt to changing conditions.

The time has come for homo sapiens to confront that eventuality. Fortunately, unlike our neighbors in the animal kingdom, we have a wide freedom of choice. We have created the very conditions that are now threatening us. And what has been chosen and created badly can be changed by the enlightened choice of we who have done the creating.

I have had the privilege of seeing our planet Earth from afar. And like a significant number of space explorers, East and West, have gained a personal sense of responsibility for its survival as a nurturing refuge for the continued evolution of intelligent life. Twentieth century human activity threatens that survival. The contemporary debate over social systems—systems of religious, political, and moral thought—while seemingly important in the short run, are but a straw in the wind of geologic time, yet they set the course for the future. Human excesses and error have historically sorted themselves out in the fullness of the centuries. But never before has the capacity existed for humans to terminate all living activity on the planet in one spasmodic hour of nuclear exchange. It is a capability existing for only a tiny fragment of time within the eighteen billion or so years of the Universe. The consequences of twentieth century human choice thus take on cosmic significance. This era must be taken very seriously as different from all those in history in this regard. The author has made his contribution to this book with the foregoing in mind.

My colleague Carol Rosin has worked with equal vision and dedication. A protégé of the late Wernher von Braun, she has devoted a number of years to his dream, augmented by her own, for the peaceful exploration and development of outer space. The final chapters on cooperative alternatives to the weaponization of space represent the contribution of her efforts.

– Dr. Edgar Mitchell

PROLOGUE

… the "rules of the game" for managing conflict must be changed when attempting to apply modern technology in the space environment.

— Dr. Edgar Mitchell

This book presents the results of four years of analysis of the Strategic Defense Initiative (SDI) and the move to effect a hasty and ill-advised implementation. The fate of the Strategic Defense Initiative is a slow and lingering death as advocates attempt to go forward and the consequences of going forward become more fully understood by Congress and the public. The arguments against and the consequences of SDI are fully presented herein. The final sections of this book suggest an alternative approach to security, one set within a positive vision of the future. That vision leads to less reliance on armed confrontation for conflict resolution than do traditional and current policies.

The result of studying SDI was alarm, significant alarm because of the self-contradictions it contained and destabilization it would produce. The conclusions suggest far greater danger to the United States and its allies, to say nothing of the danger to the world in general, than the innocuous words "strategic defense initiative" imply. There is nothing that I can find which could justify continuing SDI in any of the forms that propose to place weapons in space, now or ever. It should be completely dismantled as a weapons program. I take this position not because defense is an unappealing notion, but rather because it is a very appealing notion, but in the modern setting contains a plethora of hidden and subtle consequences that divert the objective from the one desired. The most fundamental consequence is that SDI would totally jeopardize the security it sought to enhance. Space-age technology has made traditional concepts of defense completely obsolete. Defense must be viewed with a totally new eye. Like medicines prescribed separately to cure individual diseases, when mixed, can produce a lethal brew, weapons in space produce disastrous consequences. It is the modern mix of weapons technologies and space that produces the problems. The prescribed cure is worse than the disease.

We can perhaps be thankful that this program forcefully brought to our attention an example which so vividly illustrates the wrenching problems of the modern era. Examples abound of the incompatibility of traditional thinking and modern technology. There is little likelihood that the Congress or the people would permit this program to continue after the facts were understood. Even though seldom done quickly, our processes of government eventually do find the right answer. The only danger remaining to block the correct decision is that the sufficient money, jobs, and profits have resulted from the first few years of funding to blind decision makers to the unfortunate consequences of continuing; just as those with traditional notions about defense become aware of the consequences and funding for space weapons has ceased completely.

The military and the armaments contractors tend to view this program in the same manner as all previous defense programs, just new technology and a new environment: space, to better solve the old problem of defense. But the "rules of the game" for managing conflict must be changed when attempting to apply modern technology in the space environment. The traditional rules represent pre-nuclear, pre-space thinking. The "war-peace game" as played for centuries must be played differently because it is now the spectators who consistently lose, regardless of who, if anyone, wins on the field of military contest. This has been true in certain cases in the past, but it is now true in almost all cases of warfare with modern technology and absolutely true with weapons in space. Throughout history, the social consequences of war have never had priority with military planners. In the modern milieu, they must command the highest priority if principles of self-determination, human rights, and individual freedom are to endure.

The positions of the nation's military leaders and the President as chief executive officer are unenviable. They are attempting to carry out their constitutional mission of defending the nation in an unprecedented period of rapid change and transition. Altering notions based upon five thousand years of tradition is never easy, but nevertheless the facts about warfare in the modern environment are now undeniable: 1) the resolution of international conflict by threat of nuclear arms jeopardizes the existence of the human species; 2) there is no proposed nor foreseeable weapons technology based in the space environment that does not accelerate the near term prospects for nuclear war rather than diminishing them. The first fact needs no elaboration, the second is the subject of this book.

We could expect that the checks and balances within the U.S. structure of government would arrive at these conclusions in due course and make the appropriate adjustments. If we are to survive, it must. In the case of SDI there was a "push" that attempted to pervert the normal processes. That "push" was the unprecedented efforts of the U.S. Department of Defense, set in motion by the Reagan administration, to "pre-sell" the notion to the public, the Congress, and the western world in general, and to make irreversible by future administrations the SDI effort. However noble the original objective, it would appear that the course eventually pursued has become one dominated by political and economic objectives, in spite of the emergence of overwhelming evidence that the concept is technologically and strategically flawed in all of its forms.

There is little doubt that the world at large would welcome any realistic effort to put an end to the threat of nuclear war. It is the pre-emminent threat of our century. If indeed, SDI had held any promise of accomplishing that end it could have been warmly endorsed. And if it were so, there would have been no need from the beginning to engage in the sort of "hard pre-sell" campaign of a system that would take several years, maybe several decades, to design, produce, and deploy. The schedule for eventually deciding if the program would go beyond the research phase was originally placed at some point in the early 1990s. Yet massive activity to accelerate the decision point and to assure that there would be no difficulty in obtaining approval to make SDI an operational military system began before the program was announced (see Section 4.10). In January 1987, the Secretary of Defense announced that a decision on initial partial deployment of some kinetic energy devices was imminent. Why the rush? What was going on that has not been openly told? Why were there unprecedented and misleading television advertisements to gain support for a system that was presumably still in a preliminary stage, particularly since from the beginning many analyses pointed to the undesirability of such a system? Why did retired military officers fan out all over the country and even abroad, making speeches to the public to "sell" this program? Why were the statements from the highest levels to the press, the public and Congress so contradictory, confusing and often evasive? Why were there such efforts to scuttle the only arms limitation treaty that we have, the Anti-Ballistic Missile (ABM) Treaty of 1972 (see Appendix A2).

It is well known that after military systems have progressed a certain distance down the procurement "pipeline," they are exceedingly difficult to stop, regardless of their merits or demerits. Manipulating that weakness of our U.S. system of decision making seems almost as reprehensible as the consequences of SDI itself. And it is even more reprehensible because the original notion was disastrously in error. It is a tactic worthy of totalitarian states, not of participatory democracies. Having "ends" justify "means" is a dangerous notion at any time but is totally disastrous when the "ends" are the wrong ones.

If SDI could have really lived up to the promises that launched it, that fact would have emerged as the conceptual studies and design were solidified. Like all military weapons, those for SDI needed to meet the tests of sound strategic thinking, military necessity, technical feasibility, economic and political realities. If the system could have met these tests, then indeed it should eventually be implemented to enhance the security of this country and of the world in general. If it could not meet the tests, it should have been dropped just as any other great sounding idea that does not work when carefully scrutinized. But it did not need to be "pre-sold" like a resort condominium and bulldozed through the normal checks and balances of government.

The analyses, however, have indicated that the SDI concept can meet none of these reasonable tests. The fact that many or even most of the weapon subsystems are technically feasible does not make the system or the concept valid. It is not that SDI is weak in some areas and strong in others, it fails every reasonable test of validity as a system for enhancing security. The plan can only be made to sound good if one is still fighting the last war and plans to fight the next one with more hi-tech weapons. It also sounds good if one believes that economic activity alone is sufficient justification to escalate weapons technology. If one adheres blindly to a traditional military adage, "Have more and better weapons than the opponent," then SDI can be made to sound appropriate. These were the "truths" under the traditional rules of the game that are no longer individually or collectively valid. The leap into space with weapons is not that simple. Space is a totally different environment than we are accustomed to dealing with on Earth. Our total experience in that environment spans only thirty years. Only a small number of humans have experienced it firsthand at this writing. And many of those, both East and West, have gained a significantly different view on global priorities as a result. In particular, I have had to shift my thinking as a result. What is important to do becomes different after one has seen the Earth from space. Tradition does not provide the proper lessons for launching war-making into the space environment. The brew of space flight mixed with modern weapons technology conceals a number of traps for the unwary that can be fatal.

After November of 1986, when the Iranian arms-for-hostages crisis became public, the efforts to keep SDI on track intensified. With the administration, Congress, and the public concerned with that issue, close scrutiny of SDI was lost to the more immediate need. The proposals by the Department of Defense (DOD) were designed to increase as quickly as possible the commitments to the weaponization of space in order to foreclose later slowdowns. One cannot fault the DOD for trying to do their job of keeping the nation militarily strong. We would be equally critical if they did not do so. But this time they have faulty marching orders and/or have not examined the long-range consequences of the plan. The direction is totally the wrong one for the job to be accomplished, i.e., defense of the nation. None seem to have the courage or the will to go back to the "boss" and say: "This isn't going to work, let's try something else." The problem is having the political will to recognize a mistake and rectify it after so much promotion and funding.

There is even a larger lesson in both the Iran-Contra issue, which dominated the news in late 1986 and most of 1987, and in SDI. That lesson is: in the modern era of widespread clamor for human rights and openness for democratic expression, covert operation, secrecy, manipulation, and backroom power politics are no longer viable methods for governance of a social structure.

For once we need to consider a larger and longer-range perspective before undertaking a program as massive as SDI. Decision making must not occur before the full consequences are understood. There is indeed a valid military role in space to help preserve peace for each and every country. Certainly the weapons envisioned for SDI can be developed, for there is no technological reason why they cannot be. But unilateral, unco-ordinated weapons deployment into space is self-defeating in this regard and certain to destabilize. It escalates historic military confrontations to a new "hair trigger" level.

Since this book was begun in the fall of 1986, several authoritative works have reached the public market with analyses of the technical, economic and political factors attendant to SDI. In addition there have been continuous press accounts of the debate on the subject. All public accounts which thoroughly treat the range of topics and the range of technologies involved in SDI point to similar conclusions: SDI cannot deliver what it promised; the countermeasures to a space-based defense system are simpler and less expensive than the system itself. The same technologies needed for SDI can be devastatingly used in less complex form against SDI. The additions to deterrence are non-existent. Thus the "aggressor" has an easier task than the "defender," if the defender were using SDI. This answer has been recognized from the beginning for those knowledgeable in the "rules of the game" required by the space environment. Now the studies by scientists of both East and West have confirmed this as a fact. Furthermore, the roles of "aggressor" and "defender" have become so intertwined as a result of modern technology as to be indistinguishable in the space-age scenarios, causing confusion in how to meet legitimate defense needs short of aggression.

These and many other complex issues arise. There is, however, one very hopeful possibility that has emerged from these studies: the technological war game that we humans have played throughout history may have finally reached a natural end point where the measure, countermeasure, counter-countermeasure escalation falls back upon itself. The ultimate destructiveness of nuclear weapons, the ultimate mobility of the space environment, and the ultimate speed of lasers cannot be combined without totally destroying the security that all seek. Or another way of saying it is: The modern technological environment is producing lose scenarios in the weapons escalation game. Regardless of what one does with the new technologies while playing under the old rules, everyone loses. This result requires some new and deep thought. There is a prophetic quote at the end of the cinema *War Games*: "... the only way to win is not to play."

However, in spite of these technological and economic findings, the "hard pre-sell" continued. The Reagan administration and the Department of Defense pressed on undeterred toward development and deployment of a system which they yet hope can meet some scaled down version of the original objectives. They seem to enunciate, like a "snake oil" salesman, whatever justification seems plausible and saleable to the audience of the moment. A "don't confuse me with facts" scenario is being acted out on an international stage before the eyes of the world. This can only become more embarrassing and damaging to American credibility should it continue. "Star Wars," as one scientific author puts it, "is dead!" But as to its burial, further states: "Whether that was done before the Star Wars bureaucracy has made its mischief is not yet clear. The weak links of the concept and program will eventually break, but will the victim be the American people?" The danger, of course, is that economic or political justifications will begin to dominate rationality.

On April 23, 1987, a new and thorough technical analysis of directed energy weapons (lasers and particle guns) was released by a study group of the American Physical Society. This report, "Science and Technology of Directed Energy Weapons,"[2] by a committee of prestigious scientists and engineers review of weapon systems such as the SDI, describes the capabilities, limitations, and development required for this exotic new class of weapon.

It confirms many of the conclusions of this book arrived at independently but contains less detailed and sophisticated analyses.

– Dr. Edgar Mitchell

CHAPTER 1—THE RULES OF THE GAME

1.1—THE SITUATION 1987

I looked back at Earth as Kitty Hawk approached the Moon. It was hard to imagine that only 65 years before, Orville and Wilbur were still building bicycles.

– Astronaut Edgar Mitchell, onboard Apollo 14

Following World War II, the world has experienced an expansion of technology without precedent in the history of humankind. In 1950 only dreamers, lovers, and astronomers looked at the Moon. Sixty thousand feet above the surface of the Earth was a region that only experimental aircraft, weather rockets, and balloons could attain. Few scientists and even fewer average citizens could name all the planets of the solar system, much less discuss them with authority. Space was not a subject of common or daily interest except to astronomers and cosmologists. Twelve years later we were planning to begin exploration of the Moon before 1970. That is a breathtaking rate of technological development. And the pace of basic innovations in technology spurred by economic incentive has continued unabated in the years that followed. But a problem arises in that our non-technical thinking has failed to keep pace.

Human concepts of how the world functions are still largely shaped by traditional beliefs dating into the distant past. We live in a world dominated by late twentieth century technology yet think about the human part of daily life with age-old concepts.

Numerous books dealing with the subject of runaway technology began appearing in 1968: *Limits to Growth*,[3] *Future Shock*,[4] *The Chasm Ahead*,[5] *Small is Beautiful*,[6] *Megatrends*,[7] *Seven Tomorrows*,[8] etc. This concerned thought pointed out the many dangers from the acceleration of change. But none proposed solutions that do not markedly alter the way we think and live. All require a new perspective of our institutions and our daily reality.

The human interpretation of reality derives from the system of beliefs we acquire from childhood onward through life. The sum total of such beliefs about reality is often called our "world view." "Reality" for each individual is far less determined by "how the world is" than by how we *think* it is. A truly "objective reality" is exceedingly difficult to discover even when objectivity is sought. Some threads of a "world view" are common to most human beliefs and stem from cultural traditions shaped before modern knowledge and the space-age were in existence. They hamper individual willingness to confront the future and to take responsibility for it.

The particular threads of belief which are the most dangerous to modern civilization stretch back into the distant past when humankind seemed helpless, impotent, and overwhelmed by the awesome forces around it. We still anchor many of our individual views to a belief that beyond us, out there, "they" are running the world, not "we." As a result, the level of personal commitment and responsibility for creating the circumstances of our personal and collective existence are abdicated. Perhaps before the Industrial Revolution, humans had ample reason to believe in the impotence of Man when confronted with the forces of Nature. But today, in the midst of twentieth century technology, such beliefs do not serve us well.

The solutions to the dilemma of accelerating technology require a new level of personal responsibility; commitment to knowledge and to cooperative effort not previously demanded. Thus, technological change continues to outpace our comprehension of how to manage late twentieth century civilization. Our collective "world view," our concepts of reality, are changing very slowly by comparison.

Many of the concepts about ourselves have suddenly become obsolete and often counterproductive in this technological age. Counterproductive because they lead to erroneous and damaging decisions. We will deal with several in the following pages of this book. But a few classic aphorisms that need no further explanation are listed here as examples: 1) "It's a big world out there," which is now countered with "It's a small, small world." 2) "Nature, the unending source of food and materials." It is now obvious that natural resources are finite and the ecological system upon which survival depends is delicately balanced. 3) "More is better." It has become well understood that continuous increase in every sector of human activity: population, economics, and technology, is creating problems with severe implications for the future. 4) "The universe is infinite, unchanging, and beyond man's comprehension." We know this to be completely false. And 5) "God will provide." There is no evidence that God interferes with human wrongheadedness, misinformation, and folly. It remains for succeeding generations to pick up the broken pieces of the past that human ignorance and self-indulgence create.

These and even more aphorisms of the past, though intellectually and scientifically untenable, still work their mischief on human emotions and daily affairs. As long as we humans allow ourselves to be guided by such traditional notions without critical reexamination in the light of modern knowledge, we will likely make more serious errors that future generations will need to repair, should humanity survive that long. The concern for human survival is not simply misplaced negativism. Having assumed the role of masters of the Earth and invented the technologies with which to manipulate its resources to our pleasure, we must, therefore, completely and correctly understand the consequences of what we are doing.

The biosphere that gives us life is now known to be fragile and tenuous. Likewise are the institutions that give representative government and individual freedom. The veneer of civilization is uncomfortably thin, barely concealing the underlying violence. As a result, the person on the street must assume a responsibility for being knowledgeably involved as never before in human history. The consequences of failure to do so can mean we don't survive.

Even the scientific view of how the universe functions has dramatically changed in this immediate period. Nobel laureate Ilya Prigogine,[9] in the 1970s added more evidence to end the classical notion (and hope) of Newton, Einstein, indeed all early scientific theories that the universe is a stable perpetual machine. The belief held for centuries that the universe is infinite, frictionless, and moving immutably in eternal orbits is demonstrably false. The reversibility of Nature and time, if valid in any sense at all, is only valid at the quantum level of existence. And it may not be valid even there. In other words: on the human scale of existence, and perhaps overall, the arrow of time moves in only one direction, toward the future.

The theological counterpoint that God or Consciousness is the only reality, projecting the universe toward some unknown final cause is equally contestable. The notion of an evolving planet, in an evolving universe, which exchanges energy for mass plus increased information, is more likely demonstrable. Information about ourselves and our environment, comprehended and responsibly acted upon, is the key to our species' survival.

It has become increasingly clear that if planet Earth is to have a satisfying destiny in the foreseeable future, that destiny lies in human hands. If, as foxes guarding the hen house, through instinct and tradition we consume the last of the chickens, what then? Instinct and traditional wisdom need to be augmented with new knowledge and thought before that eventuality, not after it has happened.

Because of the foregoing, it is not only important, but mandatory, that each of the beliefs by which humankind has managed itself over the centuries be critically reexamined. Even those that seem the most absolute and time tested cannot be accepted without challenge. They may likely be wrong. And in fact, many important ones are wrong in light of modern empirical knowledge. They are thus leading us down erroneous paths during a critical period in human history with our fate hanging in the balance. If even a minor portion of humankind's central beliefs about the nature of ultimate reality are found to be in error, then we need to reevaluate the human condition and the environment in which we exist before greater damage is done.

Those central beliefs about who and what we are, whence we arose, and where we are going form the basis for each person's "world view" and, in turn, a basis for life's decisions. So many elements in our contemporary belief systems were shaped by our forbearers' sense of impotence and inadequacy in front of an awesome Nature. We, the descendants, seem to cling to many of those beliefs rather than avail ourselves of more modern interpretations. Nature is indeed awesome but yields to the relentless human quest for knowledge and power—up to a point. Nature is harshly unforgiving of human folly, ignorance and willfulness, with the penalty being non-survival. Democratic institutions providing personal freedom of choice plus the powerful technology of this era impose a heavy responsibility for being correctly informed about the consequences of our actions.

There are a number of conditions that threaten the existence of life on the planet. They are all growing to crisis proportions in the time period of this and the next century. And they all stem from human activity arising out of ignorance and self-seeking. The most urgent of these are: nuclear war, unmanaged technological growth, the population explosion, and destruction of the ecological system.

This book primarily addresses the first and most immediate of the contemporary issues, the Strategic Defense Initiative (SDI), because the error of continuing it could trigger a nuclear war that renders the remainder of the problems moot. Should that error be made, there likely will not be an opportunity to concern ourselves further with the other issues. We will point out herein that SDI not only would fail to accomplish the objectives of population defense and enhanced deterrence, but that instead it would increase the likelihood of accidental war by reducing almost to zero the decision time available to avoid war. It will place a new and unnecessary "hair trigger" on nuclear confrontation that is far more dangerous than the existing condition.

A positive solution to nuclear confrontation, on the other hand, can reduce the dependence upon weapons, increase deterrence, free up resources for other pressing issues, and restore a sense of stability and sanity to modern civilization.

1.1.1—THE POWER OF TECHNOLOGY

The tools to him that can handle them.

— Thomas Carlyle/Critical and Miscellaneous Essays,
Sir Walter Scott, 1828

The risk of misuse of military technology in the past would at most destroy a tribe, a city, or even cause genocide if employed against a particular people, as was tried once in this century. The consequences of nuclear technology today are much more severe for all humankind. They force upon us the awareness that it is human choices shaped by our individual and collective beliefs about the nature of reality that require reexamination.

A beguiling fallacy that underlies thinking in our era is the omnipotence of technology. There is excessive faith that with the correct technology all problems are solvable. It leads us to attempt technological solutions to problems that have their origin in human relations. The conditions of peace and war, for example, arise from human desires, not from technology. Technology (except for information management technologies) can never help resolve such issues. Technological solutions in this case escalate inevitably to the notion that "might makes right" for the one that possesses the technology. The error of placing too much trust in technology has been an easy and natural one, for it has served humankind remarkably well during the past few centuries to eliminate much of the drudgery of life. However, we humans have used technology almost as often for our destruction as for our betterment. The problem, then, is not the technology but the choices for its employment.

We frequently animate technology as if it were one of "those-out-there" that determine human destiny, rather than a tool of our own creation which, if managed badly, can destroy us.

We humans seem to have learned little in this regard. In over fifty years of stockpiling nuclear arsenals, the awareness has increased that they must never be used. The earlier concern that nuclear war will commence from the deliberate action of the superpowers has steadily diminished. With decades of nuclear experience, the folly of a massive and deliberate nuclear exchange has become well understood by major nations, yet the buildup continues from the entrenchment of traditional beliefs and ignorance of a better solution. But the military mind set has been slow to yield to new thinking; stubbornly resisting the evidence of the consequences of nuclear exchange. Even though the likelihood of deliberate exchange diminishes, the danger lies in the escalation toward a confrontation in which the technology is so automated and finely timed that the machinery of war is loosed by unavoidable perturbations somewhere in the very complicated socio-political-technical complex. But the awareness has come *after* the threat was in existence. We were behind the problem, not ahead of it. Albert Einstein foresaw the dilemma near the end of his life and was quoted:

> "The unleashed power of the atom has changed everything save our modes of thinking, and we thus drift toward unparalleled catastrophes … Our defense is not in armaments, nor in science, nor in going underground. Our defense is in law and order."

> – Reported by Ralph Lap, "The Einstein Letter That Started It All," *New York Times Magazine*, August 2, 1964

In the following sections, we will continue by examining certain traditional thinking that is relevant to the SDI problem, then the technologies which were proposed for SDI to enhance deterrence, and finally alternative approaches to security. In order to see the issues in their proper perspective, the reader is urged to view the matters we are addressing in the way that an extraterrestrial observer might look at the unfolding scenario on planet Earth:

Imagine Earth as a globe you can hold in your hand. This perspective causes one to see the problems from a much broader viewpoint. The beauty and fragileness of our biosystem is better understood when it appears in this way. The foibles of humans, while still ludicrous, are less personal from that vantage point. The figures of history are but actors on the stage, even though our ancestors. There are just a few questions in the mind of such observers: "Is planet Earth likely to resolve its nuclear weapons crisis, given its history of warlike behavior?" "Is Earth likely to have a future history if the current patterns of conflicting and archaic beliefs continue?" "If so, what is that future?" "If not, what are the possible alternative choices that might produce an acceptable future?"

1.2—THE THEORY OF GAMES

*There is a saying in the world of public communications that if
you can't dazzle them with brilliance, baffle them with bullshit.
Since I can do neither, I will attempt to be articulate.*

— Dr. Edgar Mitchell

In the 1940s, a brilliant mathematician, John von Neumann, developed the
Theory of Games.[18] It was his contention that when and if all the rules
of a game and its environment are described in detail, then the optimal
strategies for winning could be discovered. All games in principle have
such strategies, he contended. But when this happens, the game, in effect,
becomes trivial and without interest. With such an approach, computers
can learn to play chess, for example, by searching all the possible strategies. The only really interesting game is then the "game of life," where
the rules can change, people bluff, lie, cheat and break the rules. Not only
do the rules change, but the environment is not completely or accurately
known. Thus the outcomes of strategies in the game of life are not always
certain, but they can be studied as to how a change in strategy changes
the likelihood of an outcome. Thus "game" and "reality" can almost be
used synonymously. Since those beginnings, game theory concepts have
been carried into the computer world, into psychology, into management
practice, into the fields of motivation and personal development, to name
a few, in order to help us better manage our lives. But seemingly, they have
never been introduced into the highest halls of government.

In the modern world of computers it is usually very productive to attempt to simulate complex problems on a computer and to think of them in game theory terms. In this way one can test various outcomes as the strategies are changed. I have not been required to use computer simulations in arriving at the results presented in this book because the principle outcomes from technological change can be discovered with rather simple parametric and economic analyses. The results of the SDI study are not sensitive to which of the new generation of weapons is employed in the "war games." They all produce movement toward the same outcome. The results are sensitive, however, to small changes in political policies because it is the psychodynamics of human decisions to escalate weapons that actually drives the outcome. I will describe these situations with simple human scenarios. I will demonstrate in the following sections that the historical escalation of weapons technology is the inevitable result of the strategies historically pursued. It is descriptive to borrow some terms from game theory to help illustrate the conclusions. Most people should be familiar with the ideas and terminology. To illustrate:

In any game with two players (or two sides) the possible outcomes are four: win-win, win-lose, lose-win, lose-lose. Regardless of the complexity of the rules or the situation, the contest between two sides can only have four outcomes. One normally only considers win, lose, or tie as the outcomes in a game, but in actuality there are four. Certainly there are distinctly four if life or death is the possible outcomes for each player. A tie can be of two sorts: one where both players live or both players die, thus win-win or lose-lose. The terms seem very descriptive and need little elucidation. If we call the outcomes of the war-peace game: live-live, live-die, die-live and die-die, there certainly should be no mistaking what we mean by each player's outcome in a war game.

We will refer frequently to the "rules of the game." The "old" rules of the game are those historically in use for our personal and national relationships. They represent the idea of "business as usual" with very little thought as to long-term consequences. The "new" rules of the game, by contrast, descend from a world view that recognizes the finiteness and fragility of the planet. They recognize that ultimately there is no security for anyone unless there is security for all; that the biosphere which nurtures life on Earth is a fragile interactive system and in the modern era, easily disrupted by human activity; that evolving human life and human activity are growing in ways that threaten our very existence as a species.

The rules of the game are what we humans choose in order to govern our activities. Of course, one can always choose not to play the game. That is another game called "Stop the world, I want to get off."

In playing the game, before the end is in sight the outcomes can only be expressed in terms of probable outcomes, i.e., probabilities. The sum of the probabilities must equal 1.0, since there is always an eventual outcome if the game continues to its end. As the numerical value of a probable outcome approaches 1.0, that means it becomes certain. If a probability is zero, it means that the outcome will not happen. The game of life is to "beat the odds," but it is difficult to beat the odds when a particular outcome is certain, that is, has a probability of 1.0. Readers who enjoy placing a wager now and then should have no difficulty with this idea. The probabilities just express the "odds" that a particular outcome will occur. Instead of expressing the "odds" in the ways of the betting world, we are expressing the odds of an outcome on a scale from 0.0 to 1.0, with 0.0 representing the "long shot" with a low probability of happening, and 1.0 representing the "favorite" that is a "sure thing."

Throughout history military men have always thought in terms of only one outcome, the win-lose outcome: we win-they lose. All strategies are shaped in that direction. Clearly the other side must be in the position of lose-win with such strategies (if "we" are successful, "they" cannot be). We hope to demonstrate that traditional strategies shaped toward win-lose solutions are often not viable in the modern era because they lead to lose-lose outcomes when modern technology and space are combined, not on the computer but in real life! Strategies that traditionally have led to a probability of winning now cause losing, not for one side but for both; most particularly for the populations. In Viet Nam, for example, when it became clear that the win-lose strategy had resulted in a zero probability for winning, we stopped the game to keep from losing. But of course many had already lost.

We will use the term "population" to refer to the people and to their assets. Their assets being the national assets: homes, cities, farms, factories, the means by which we humans develop our private and economic lives within our respective societies. We will use the term "champion" to refer to the strategic military forces, military leadership, and military assets whose function is to defend the population, the state, and national assets. The "champion's" assets of most concern in this book are the strategic nuclear arsenals and their attendant systems, including the new generation of space weapons being developed. Conventional military arsenals and strategies will be specified as such in the final sections of the book where new solutions to the problem of conflict are suggested.

It might be appropriate for this book to make a distinction between the professional military and the civilian political leadership which directs the military. Quite often it is an important distinction. Because, however, we are contesting ideas and not individuals, we will use the broad brush and consider as "champions" the entire strategic military establishment, uniformed and civilian. There is full recognition that there are individuals in both groups that agree with the arguments herein and many of both groups that disagree. Traditionally it is the political leadership that must hold a restraining hand on the military desire for even more modern weapons. We tend to see a reversal in this case where the political leadership is rushing toward an end point for which large numbers of professional military express great concern. It is that reversal that prompts this effort.

Much of the change that must be made in human thinking involves what I call the "traditional military mind set." It is as old as humanity itself. It is thinking hardly confined to military persons but captures all of us when belligerence and hostile attitudes cause eagerness to fight before all other solutions to conflict have been attempted. One can epitomize the notion with a phrase attributed to General George Patton, the magnificent warrior: "War is hell—but, god, I love it!"

We also use the terms: first strike, first strike capability, and preemptive strike. These all seem to be referring to the same thing, but they are not. In current military usage, first strike capability refers to the ability to target a strike such that effective retaliation is precluded. This implies the ability to guide nuclear missiles close enough to a hardened silo or command post target to have some designated probability of destroying it; similarly the ability to locate an underwater submarine with sufficient accuracy to do likewise. The term may be thought of in the sense of individual targets, in which case guidance accuracy is the main element, or it may be thought of in the broader sense of neutralizing all opposing targets by such means. The latter implies a strategic doctrine in addition to a technical capability. We will discuss these notions in more detail in section 4.4, First Strike and Retaliation. First strike can refer to the pursuit of such capability but can also mean just what it says—to strike first. Preemptive strike is used to mean a first strike when the opposition is believed to be planning a strike. This is the normal connotation of the phrase.

The guidance accuracy of weapons is described by the "circular error probable" (CEP) which is the median radius of expected hits. It is an empirically determined measure of accuracy. Fifty percent of impacts will fall within a designated CEP and fifty percent outside. Thus a weapon with a guidance accuracy (CEP) of 500 feet would mean the explosive is expected to land within 500 feet of the target point fifty percent of the time.

There is no legally accepted definition of space. A generally accepted notion is "that region where unpowered orbital flight is possible." This would be the region above 150 kilometers from the surface. Air breathing engines are effectively limited to below 30 kilometers, thus there is a significant area above the zone for aircraft and below orbital altitudes which is not easily used for human activity. Undoubtedly, a legal convention will eventually be adopted to define the space boundary and with it the extent of a nation's sovereignty over the airspace above its surface area.

Likewise weapons for use in space do not have an accepted definition. For this book I will construe weapons to mean "active weapons" capable of directly destroying, damaging, or interrupting the function of material objects. Although it is a point for debate, I will not interpret devices such as sensors and radars to be weapons even though some such devices would need to be used to control and direct active weapons.

1.2.1— THE NEIGHBORS

Any man more right than his neighbors, constitutes a majority of one.

— Henry David Thoreau, *Civil Disobedience,* 1843

Warfare, like love and hate, is never really rational. War is so interwoven with human passions, traditions, needs, and beliefs, that rationality is usually not the determining factor at all. Let's examine a simple human scenario that has obvious relevance to the current state of international conflict. I want to demonstrate in a simple example the human processes that develop. And to show the relationships that arise among intent, deterrence, mutually assured destruction, and technologies. After all, it is not governments that make war or peace, it is particular individuals within governments who make the decisions that lead to war. The nebulous and never present "they" that cause things to happen are in reality specific human beings that must accept responsibility for their actions.

Let's assume that one has an old World War I revolver hidden away from the children in a bedroom closet. It has been there for years as a memento but also "just in case it is needed" to defend one's household from an intruder. It clearly is a weapon that might be used for one's own personal defense. The pistol hardly even qualifies as a defensive weapon because it has been there so long and is almost forgotten. It is just an old souvenir.

But that very same revolver, ancient and rusty, if loaded and held against a neighbor's head in his house, and not one's own, has suddenly become a weapon for offense. The weapon hasn't changed but clearly the behavior of its possessor has. The individual has changed from defender to aggressor, the location has changed from one's own territory to that of the opponent, but the weapon is the same. And clearly the intent has changed.

Let's now examine some of the steps that lie between one being a peaceful, good neighbor with an old revolver stored in the closet "just in case" and one being an aggressive neighbor with the old revolver aimed at the neighbor's head, ready to pull the trigger. Understanding this problem in psychology is as important to the wellbeing of twentieth century humans as is understanding the technology of warfare. Maybe it is more important if the issues are phrased in terms of moral righteousness and the outcome of the conflict could destroy the entire neighborhood. Governments don't have minds or need psychology, but people do.

Let's assume your neighbor is an unpleasant type with whom it is easy to disagree. Muttered obscenities are difficult to suppress after even a casual encounter. Sometimes tempers flare and harsh words are exchanged across the back fence. Maybe even some veiled threats are made by one or both parties even though sincerely regretted later. But the neighbor's intentions do not appear to be honorable or honest, or even safe. The old revolver in the closet is remembered. So, on impulse you take it out for inspection just in case that crazy fellow next door starts to attempt something stupid. He appears capable of such foolishness, and you have a responsibility to wife and children.

If the neighbor has no knowledge of the revolver and you have no intention of using it unless attacked in your own household, then no harm is done. The revolver remains in the closet and is again forgotten. It is just a souvenir. But what if by chance or design, the neighbor saw the pistol being inspected after the disagreement? The shades were not drawn over the window. What is he likely to think? He is likely to assume that he is in danger and needs to look out for an attack. He might call the police. But if he is fiercely independent or there are no police to call, then he is likely to make sure he has a weapon himself. Now we have two defensive weapons, one in each household. And if each is aware of the other's weapon, they likely will not be kept in the closet, but more easily accessible—cleaned and ready for use.

If both of the neighbors, regardless of their mutual dislike, are relatively level-headed fellows, things can stop at this point and the two can coexist in the same neighborhood without too much difficulty by avoiding each other. They live with mutual dislike. But what if by chance they are in competing businesses? They are trying to sell their companies' products (or ideas) to the same clients. If business is good and each is making sufficient money (or converts), the strained but peaceful coexistence may continue. But business is not always good; markets have a way of becoming soft from time-to-time. Then the antagonism and irritation increase. They may even be of different religious persuasions and ethnic backgrounds, or worse, different politics. That will certainly make the situation even more tense. There is likely little understanding, little respect, or agreement on any subject. Each will certainly begin to see the other as an aggressive, ruthless, dangerous opponent.

Let's suppose that you like to hunt fowl and for that purpose you eagerly buy a new shotgun for the duck season. But the suspicious neighbor sees you bring it home. Despite the fact that shotguns are normally used for bird hunting, the neighbor is now thoroughly alarmed at the thought of additional weapons next door in the hands of his unfriendly neighbor (you). He places extra locks on the door and heavy shutters on the windows "just in case." And certainly if he is such a frightened man, very shortly a new shotgun appears because some rationalization makes it seem important. Now there are two pistols and two shotguns. But you know he doesn't hunt. Hunters are fine gentlemen and sportsmen; this neighbor is just "sick." Soon each has a stick of dynamite beside the guns.

All that is now required is for each to have an automatic catapult to retaliate by throwing the sticks of dynamite across the fence if a foreign object (including the neighborhood cat) comes flying over from the other side. Mutually assured destruction (MAD) of each other, and perhaps the neighborhood, is at hand. It is only a matter of one or the other becoming fearful, desperate, or enraged enough to shoot first (or the cat going for a stroll over the fence). In this case it's all over for both, plus others, in a matter of 15 or 20 seconds. Which one is the aggressor, and which one is really peaceful is moot. The motivation of each was to deter the other's aggression. Their policy was deterrence. The result was the mutually assured destruction, not only of the two neighbors, but of their families as well.

The scenario is not trivial regarding nations. Certainly the Mid-East scenario of today is but a larger variation on this theme, exacerbated by population growth, thousands of years of mistrust, and couched in phrases of moral and religious rightness. And it is the population that is taking the real losses. With modern technology, such disagreements can only lead in this direction, unless one or both neighbors make a more level-headed assessment of the situation. Fortunately, most of us do believe we are civilized. The underlying problem is fear. Having strange neighbors, particularly if they are conspicuously armed, leads to a consuming fear. Without a legal jurisdiction to which both are subject for the settlement of disputes and maintenance of the peace and a willingness to submit to that jurisdiction, someone will eventually get hurt. The weapons each possesses have never been the real source of the problem, they are just a means of implementing what one has in mind—that is, until they become highly automated. They pose a totally different order of problem when automation technology is introduced. But it is the attitude and the intentions of the neighbors that first determine how the weapons are employed. Furthermore, it makes no difference if both believe that they themselves are only armed for defense, it is the belief about the other person's intention that causes the escalating reaction.

No other single idea in this book is more important than this, so I shall repeat it: It is not what you believe about your own intentions, but what you believe about the other person's intentions that determines "defensive" behavior. And since one seldom knows exactly what is in the mind of a friend, much less in the mind of an enemy, it is fear and uncertainty that causes mistakes. In the case of an enemy, one is inclined to believe the worst. With a little effort we can convince ourselves to believe anything we choose to believe, particularly concerning our own rightness and the opponent's wrongness. This is always more likely if we are fearful and uncertain of what a well-armed opponent might do. The interaction of intention, belief, fear, uncertainty, and, finally, technology, are the real variables in the drama. These are the variables examined throughout this book in the actual nuclear confrontation scenario.

Observe what happened when the neighbors introduced the automatic catapult into their quarrel in order to shoot back immediately if the other attacked. Before the automatic machinery to launch the explosives, the probable outcomes in the situation were totally determined by the attitudes of the two parties. They were still personally in control of the situation. After introducing the automatic control technology, the likely outcome changes. In the manually controlled "war," the probabilities of each of the four possible outcomes were perhaps equally likely. But whatever their numeric value, they were determined by the opponents' attitudes. The introduction of the high-explosive dynamite caused lose-lose to increase as the likely outcome, since one explosion could very well kill both. The introduction of the automatic catapult creates strong additional bias toward the lose-lose outcome without any change in opponent's attitudes. The probability of win-win, win-lose, lose-win equally shrink toward zero and the lose-lose outcome dominates, increasing toward certainty. Automation alone has increased the probability toward a certainty that both will die. Even the cat going for a stroll could initiate the accident and become the most likely trigger. Their only options are to live with the situation or dismantle the automatic catapults. They have no others.

In the example, as in the real case, neither side deliberately developed a policy of mutually assured destruction. They were trying to create deterrence by being sufficiently armed to preclude the unpleasant neighbor from attacking. The policy was deterrence, but the result was MAD if deterrence failed, or if the cat went over the fence. The cat represents the "accidental" triggering of the lose-lose outcome. Accident becomes a major factor as automation increases and warning time decreases. If people are rational, a condition whereby mutually assured destruction arises can be a stable situation, provided both sides can learn to live with fear and uncertainty. In the actual situation, we have lived with the condition for about thirty years. It is moot that getting into the situation in the first place appears irrational. We humans get ourselves into impossible situations almost daily and before we recognize it. It is also irrelevant, after the fact, that the policy creating the condition was deterrence; the probable outcome if shooting begins for any reason is MAD. Thus we distinguish between the policy's deterrence and the condition which the policy produced: mutually assured destruction. Enhanced technology makes it more certainly a lose-lose outcome and puts a hair trigger on the machinery.

The condition of MAD, even though stable, produces fear. No one can argue that it is an undesirable condition. Fear enhanced by uncertainty is the driving force for erratic and irrational behavior. One can make a strong case that reaction to fear is the driving force behind much of our socio-political decision making rather than rational thinking. The more tense and active one side becomes, the more the fear is increased on the other side. Thus the more likely is the perpetuation of erratic behavior. We must assess that, at best, a MAD condition arising from conflict can only be marginally and temporarily stable. It can become unstable and provoke more fear to the point of desperation with the smallest of perturbations. The more aggressive or fearful are the opponents, the more dangerous and unstable is the policy of deterrence by threat of retaliation. Once attained, the condition must be handled with utmost delicacy to avoid precipitating an instability which exacerbates the condition. Fear, uncertainty, a nervous constituency, "psych" passing as justification, and warning times reducing toward zero are clearly instabilities that increase the danger, not reduce it. The warning time problem is particularly precarious.

Notice that all of this can occur even though the intention for deliberate attack does not exist. If, in addition, either opponent holds the idea that selective use of his weapons can disarm the other, then the probability of deliberate attack is not zero and the likelihood of activating the war process is increased.

1.3—THE MYTH OF DEFENSE

I have often been asked if we carried weapons to the Moon in the event extraterrestrials were there first. I reply that if they were there first, they would be more advanced than we, thus we wouldn't need them. (And omitting that that likelihood is virtually zero.)

— Dr. Edgar Mitchell

A folk adage that has descended from some unknown source of the past is: "The best defense is a good offense." And certainly in many situations that would seem to be a valid aphorism. But civilized people can never feel comfortable using such a notion as the central dictum of their military policy. Those that do use it are little better than barbarians rationalizing their aggression. Such notions are best reserved for the playing fields of schoolboys. Modern humans who prize freedom, human rights, and equality cannot equate defense with offense except as a measure of last resort.

Escalation of weapons technology in the contemporary era, however, has rapidly forced all nations confronting an equally powerful adversary to consider offensive capability as their defense. We will argue herein that the natural consequence of nuclear technology, plus the technologies envisioned for space use, not only make the adage absolutely true but, further, make first use of the technology mandatory if one believes defensive action is necessary.

Specifically, with space-based weapons of mass destruction (i.e., lasers and nuclear weapons under control of an opponent), preemptive strike would become the only defense in time of crisis. The inclination to be passive or peaceful would be overshadowed by the necessity of an aggressive preemptive strike in the presence of such weapons. In this situation, humankind's civilizing desires become slave to the technology causing our traditional notions about offense and defense to be muddled and misleading.

I am not suggesting that aggressive intent does not exist, merely that the traditional roles of aggressor and defender become hopelessly confused because of the technology. I will develop several situations to illustrate this problem.

Throughout all history peaceful people have needed adequate defense. And warlike people have needed defense even more because there was always some victimized group looking for revenge. However civilized we become, attacks upon person, possessions, or even strongly held beliefs raise visions of revenge. It is not difficult to establish the case that from the beginning of history adequate defense has been a realistic and necessary concern of people everywhere. It is likewise not difficult to reach agreement that human defensive security has always resided in four principal factors: superior numbers, superior firepower (even sticks and stones), superior knowledge, and a defendable position. These basic elements of defense have never changed in all time—until this century. Only the technology of firepower and the tactics for using it have changed. The principles have not.

A defendable position out of reach of predators and adversaries has historically been the major concern of all people trying to assure their safety. A cave, a steep mountain, a walled city, a double walled city, moats around the walled city, a castle built upon a steep hill, fixed gun emplacements, fortified cities, perimeter defensive outposts, the Maginot Line, represent the progression of attempts to have a defendable location in which to live. The less naturally defendable the terrain, the more artificial the defenses have become. The technology and the strategy for defense have necessarily become more sophisticated with improvements in firepower of the aggressor. The Maginot Line built in this century to defend France against Germany was, however, the last of the great defensive systems to rely upon fixed positions and fixed weaponry. It was totally ineffective against airpower and the high mobility of modern aggressors. The Maginot Line provided, in fact, no defense at all. The invader simply went around it and over it with impunity. With the Maginot Line, strictly defensive systems disappeared.

On the other hand, what are the factors necessary for a successful aggressor? There should be little disagreement that they are: superior numbers, superior firepower, superior knowledge, mobility, and if possible, surprise. These factors likewise have not changed in all recorded history. The only difference, then, between the requirements of defender and aggressor have been that a defender needs a defensible position, and the aggressor needs mobility and, if possible, the element of surprise.

The history of warfare has centered around improving the technologies associated with firepower, gaining knowledge, mobility, and defensible positions. The first modification of the factors which distinguished the needs of aggressors from those of defenders came as newly acquired territory required defense. The prized territory for habitation is open, fertile fields which are not easily defendable. The best defense then requires mobility of firepower and artificial defenses to counter the aggressor at the border. Mobility, beginning as a need for aggressors only, quickly became a need for even "peace loving defenders" when their home territory became extended through migration or aggression. With this evolutionary modification of the factors, the only distinction between defenders and aggressors then becomes: defendable location on the part of defenders and the element of surprise on the part of aggressors. Otherwise, the elemental needs for defense and aggression are identical. This situation has prevailed for the largest part of recorded history.

The fact of defenders acquiring mobility, however, makes the defender look capable of aggression to everyone but himself. The technologies of firepower, gaining knowledge, and battlefield mobility have, of necessity, been developed equally by defenders and aggressors alike. The difference has often been only in the quantity of arms and the willingness to use them. The need to be "superior" in order to adequately defend or to conquer has assured escalation of technology. Wide mobility was developed as a result of trying to defend ever larger territories and to assist allies in their defense. The oceans, the skies, and finally space, have provided the means for achieving that mobility apart from the land itself. The oceans, the skies, and space, unlike the land, however, are not defendable locations. To view them as such invites error in strategy. Each has required technological advancement in order to be utilized and each is progressively less defendable. They can only be realistically considered as a means of achieving mobility.

The final and decisive change in this scenario has come late in the twentieth century. Until the current era, indeed until the past four decades, being a defender required merely being prepared to defend one's home territory or that of an ally. Distant people have never before been threatened, until the utilization of space for mobility by other distant peoples' defensive preparations. Movement of men and machines over great distances on land, sea, or air has always brought the equally high price and difficulties of logistics with long supply lines. Surprise attack on a well-prepared but distant defender has never been a very feasible nor intelligent strategy for any aggressor bent on conquest. Many conquerors down through history have tried it, however. Those that succeeded, such as Alexander the Great[10] and Julius Caesar,[11] were hailed as having military genius. Only the hope of significant economic gain makes the risk worthwhile; usually only for the foolhardy. Unless the defender is very weak, conquest over long distances seldom succeeds. But, in fact, every major power in the world today has had their period of conquest and acquisition against weaker defenders, beginning, in the modern era with the colonization period in the fifteenth century.

Aircraft and intercontinental missiles have now changed the equation completely. Whereas previous defenders needed only short-range weapons to defend against the aggressor's attack, modern technology has allowed the attack to be launched from the aggressor's home territory anywhere in the world and without re-supply. But the "defensive" weapons must be able to counter with equal range. They look identical to the aggressor's weapons. With short-range weapons, limited mobility, and the need to massively transport manpower, it was still possible to make a rather sharp line of distinction between the aggressor's military needs and those of a defender. The aggressor's intentions were obvious if military buildup for offensive action had begun, since mobilizing a campaign cannot be done quickly nor without giving off telltale signs. Defenders could rely upon good intelligence to alert them in sufficient time to arm.

With the new long-range weapons of mass destruction, the next to final difference between defenders and aggressors has vanished: defendable location. There are no longer any areas of the world that are "defendable locations." "Defensive" aircraft and missiles can make strikes anywhere on or above the surface. They are, in fact, offensive weapons. The only safe locations are deep underground and, in some cases, deep underwater. But unlike homes and land, those are not locations that the public values as a "defendable location." Defending one's home, family, and means of livelihood is instinctive, but defending remote underground and underwater sites has no personal meaning whatsoever. The areas of ocean, sky, and space immediately adjacent to one's territory may be considered as an adjunct of one's domain, but nevertheless, they are primarily for mobility and not naturally defendable.

With the historic progress in technological innovation, humans rapidly became the most vulnerable part of the battle. An unprotected human is too easily put out of action by the firepower of ever more modern weapons. From face-to-face confrontation with spears to push-button warfare, history has recorded slow but continuous movement toward replacing superior numbers in manpower with superior numbers in firepower and in the sophistication of weapons. The rapid part of this change has come only in this century. It attests to how recent in the past, however, that the dress uniforms of many military organizations still included a sword, but swords haven't been used in combat in this century!

In the current era, with missiles, nuclear weapons, unmanned space craft, and computers, the soldier is the technician who operates them and is far from the battlefield. Technicians and the officers in command are buried under tons of protective concrete or tons of sea water. Only the conventional military in conventional and special forces warfare is still at the front lines and in tactical confrontation.

Another important difference has emerged in this century to confuse the roles of aggressor and defender and to give reason for reevaluation. Nuclear technology, if used in large amounts as would be the unavoidable case in any war between major powers, renders the land useless for unforeseeable periods. The land could not be immediately occupied and made productive. Thus, an aggressor is thwarted from using his nuclear firepower for conquest. He must find other ways of exercising and enforcing aggressive ambitions. The offensive nuclear firepower is thereby reduced to a defensive retaliatory threat; and with that, the positions of aggressor and defender become merged and indistinguishable except for the intent to surprise. The use of nuclear firepower leads to wanton destruction of the world's desirable assets and is thus unthinkable by rational persons, be they aggressive or defensive. Its only use would be an irrational act of desperation to avoid defeat but would even then create an escalated counter-use leading to total destruction of both sides. Mutual use thereby produces the lose-lose outcome of game theory. And in terms of the Nuclear Winter Study,[12] massive use by only one side is sufficient to bring on a lose-lose outcome for both.

To underscore this issue: In March 1987, a scientific paper was released announcing findings which suggest that the Earth's Ice Age may possibly have been triggered by an asteroid impact on the Earth. The impact, two and one half million years ago, was estimated as the equivalent of 172 fifty-eight megaton nuclear explosives, the largest ever tested. (New York Times, March 16, 1987) The nuclear arsenals of the superpowers together contained the same order of nuclear explosive potential in the 1980s, about 10,000 megatons. Thus, the evidence steadily mounts that the biosphere's capacity to withstand nuclear assault without irreparable damage is but a fraction of the existing nuclear capability. If humans understand this fact, deliberate (i.e., rational) use of nuclear weapons, should have a near zero probability of occurring.

It is interesting to note that the near simultaneous development of nuclear weapons and of intercontinental missiles produced offsetting results. Nuclear weapons may be the ultimate weapon of destruction. Certainly space seems to provide ultimate mobility for an aggressor. Combining the two with both adversaries so armed reduces the combination to a retaliatory threat which, if used in quantity, produces lose-lose outcomes.

Thus we arrive at the point in the modern age where the single remaining military difference between defenders and aggressors is surprise. Otherwise the technologies, the mobility, the knowledge, the problems, are exactly the same for each. One cannot look at the arsenals of the major powers nor at their defendable locations and make any distinction between aggressor and defender nations. They look exactly the same. With missile flight times of only 30 minutes between the superpowers, theoretical warning times against surprise attack are reduced to 30 minutes as well, even in times of peace. Defenders must be ready on 30 minutes notice, tantamount to being perpetually prepared, the same as an aggressor.

Technology thus has become supreme and dictates the conditions of readiness. The fact of possessing the weapons requires constant preparedness to use them on short notice. The only difference remaining between the opponents is in one or the other's intention to use their weapons. Any shift in the readiness posture of either side, which the other minutely monitors and can accurately discern, is taken as a possible intent to attack. The best defense is completely centered around the need for an identical best offense. Surprise, the aggressor's only remaining element, has been compromised. Surprise, first strike, and preemptive strike are concepts now used by defenders and aggressors alike to assure the adequacy of their defense. The technology requires it. Surprise and first strike, which have always been the final test to mark the aggressor as being "the aggressor," are now of necessity being considered as a defensive option. In this condition, warning time becomes a most critical element. It becomes, in fact, the determining element in the nuclear war game scenario. With warning times approaching zero, there become no other options available to a "defender" who believes that an "aggressor" intends to strike. It might not seem that a difference of only thirty seconds in deciding to begin a war could very well be significant, but as I shall demonstrate, it could be the most important thirty seconds in the history of the world.

With such concepts being used on both sides of a conflict, the last imaginable distinction between aggressive activity and defensive activity vanishes. Only the intent remains to distinguish the opponents. But if aggressors and defenders are indistinguishable except by their intent, then even legitimate defense needs could be easily and most likely interpreted as aggressive intent. All military activity seems like defense to each "defender" and must be viewed as aggressive intent by the other side. In actuality, neither defender nor aggressor requires justification for the need to defend, it is instinctive and necessary for all. An aggressor, however, must have at least some semblance of a rationale to justify his actions to himself and his constituency. And this rationale provides clues to his "intent."

We see that "Defense" plus "social or moral rightness" are emerging as the rhetoric to justify increased military activity on both sides. This is true regardless of the opponent to which one listens. With there now being no discernible difference between the military activities, nor the rationale or rhetoric of aggressor and defender, an observer cannot determine which is which by any objective test. Only subjective measures remain. In such a case the activity itself becomes the danger—not who is doing it or their reasons. Technology becomes the master, not the slave. The need for "superiority" on both sides drives an unending escalation. All that is then needed to create a war which no one can win is for one defender or the other to *perceive* that a preemptive strike is necessary for his own defense, or for one to misjudge the moves of the other. There is no requirement to have aggressive intent to ignite the explosion in this case, just a defender's legitimate need to defend. This will surely happen as both sides escalate into space with weapons that cause warning times to approach zero. Lose-lose becomes a reality, not a computer simulation.

We will examine in a subsequent section the case that would likely develop from the SDI use of space technology. The real crisis arises when the warning time approaches zero, ultimate surprise is possible, and the decision time for counterattack is near zero. Preemptive strike then becomes the most likely defensive option, and/or accidental strike takes place because decision time is insufficient to prevent it.

1.3.1—COUNTERPOINTS

There is no error so monstrous that it fails to find defenders among the ablest men.

— Lord Acton, Letters to Mary Gladstone, 1881

Remarks by Reagan's Secretary of Defense, Caspar Weinberger, on the "arms race syndrome" would seem to dismiss the foregoing points:

"For example, one of the most pervasive myths about our relations with Moscow concerns the so-called 'Arms Race' syndrome … According to this view, the United States and Soviet Union are really 'Two Scorpions in a Bottle,' or in the catchy phrase of a former arms negotiator, we are 'apes on a treadmill.' Thus, Washington and Moscow are two equivalent powers, driven by an obsessive anxiety about security and reacting rather mindlessly to each other's arms programs. Moreover, the two share the same strategic doctrines, both build essentially the same kind of weapons, and both equally menace world peace … The fault with this kind of thinking is that reality keeps imposing itself on the idea of political and moral equivalence."

— U.S., Office of Assistant Secretary of Defense, News Release: "SDI and the American Strategic Perspective," 26 September 1986

Having heard this, one is prompted to ask a question regarding the fore-going quote: What "reality" imposes itself on whose idea of political and moral equivalence? Morality is certainly important to human wellbeing, but it remains a subjective value judgment. Reality, I would hope, has a bit more objective flavor. Discussing morality and reality is this context has no real meaning. Surely, we are not preparing to wage a global war over ideology as in the Middle Ages or as they are claiming to do in the Middle East. Is the Inquisition still with us to force adherence to some particular notion of morality through military action? Or is it that type of thinking and rhetoric which drives the entire dynamic system of super-power conflict into a cul-de-sac, within which there is no alternative but for one or the other "defender" to launch a preemptive strike that acti-vates all the machinery of war on both sides? It would seem that a more reasonable interpretation of the current problem is that the reality of military and political equivalence is offensive to Mr. Weinberger's idea of morality. Rationalization of moral rightness becomes ridiculous when the legitimate activity of defense itself escalates to become the problem. The claims of defensive need or moral rightness sound more like "psyching" the team up for the battle.

"Psyching" ourselves up for the battle is important if it is certain there will be a battle. It is perilous if one is attempting to avoid a battle, since battles overzealously prepared for are the ones most likely to be fought. Being too ready to fight usually assures the fight will occur. This idea is absolutely vital to understanding the dangers inherent in the SDI buildup. By focusing on the rightness of only one position and being "psyched" toward preparedness and instant response, the danger is sure to increase. Add to this, technologies that reduce warning time toward zero so that any movement of the opponent must be immediately construed as aggressive intent, and the stage for accidental conflict is set. One is not required to like the opponent, his social structure, nor his morality, in order to comprehend his legitimate needs and his likely responses. In the current highly controversial scenario surrounding SDI, movements in either op-ponent's strategic posture, regardless of actual intent, can only be viewed by the other as indicative of aggressive intent or irrationality, with the latter being more dangerous. There is sufficient activity on both sides for each to reasonably believe in the aggressive intent of the other. It would seem that an in-depth knowledge of psychology would be more import-ant for a Secretary of Defense than knowledge of military technology.

The "apes on a treadmill" analogy is only appropriate if we continue to react mindlessly to the old rules, instead of as creative humans who can change the rules when it is in our mutual best interest to do so.

An additional fact was buried in the foregoing that might have been overlooked: with the "missile warrior" no longer being at the battle front, but instead operating the remote controls of battle from heavily bunkered positions underground, underwater, or at a remote location in the air away from impacts, the warrior is less at risk in a nuclear or an orbital war than is the population he is defending. For the first time in history, the people's "champion" is less at risk than the people during a war. This is a significant fact. It is a fact which dictates in part the necessity for new "rules of the game" in the modern era of missile and space warfare. It requires a change in priority for the decision-making process. A relatively few men at computer terminals and connected through communications links control the entire system of battle management. And they are located in the midst of the population. This fact forces one to compute separately the probable outcomes of war for the populations and their champions since they are no longer the same for the two groups. A champion's win does not assure that his population has won also.

From the foregoing we may conclude that, whereas in the past it was possible to distinguish between peaceful and aggressor groups by their military posture and their behavior, it is now impossible to do so. One must now ask the questions: if the defender in defense of his territory launches a preemptive first strike, is he still a defender? Is one's defense of moral righteousness still righteous if all existence is threatened? Or does it really matter what one calls it? By whatever one calls it, a rose is still a rose and Armageddon is still a war of annihilation. The first needs no explanation and smells just as sweet; the second cannot be explained if no one remains to hear. The danger is that in spite of human rhetoric and explanations, the imperatives created by the technology and guided by "moral righteousness" may preclude rational decision making. The story of the sorcerer's apprentice can be acted out, but in this case, there is no sorcerer to set things right.

1.4—MORE ABOUT DETERRENCE

I once was confronted by a man with homicidal intentions.
Looking at his eyes, I discovered he was more frightened than I.

– Dr. Edgar Mitchell

In the preceding section we introduced the ideas of deterrence and mutually assured destruction. The latter describes the condition that has arisen as a result of the high technology arsenals possessed today. It is a popular term, not an official term, but is accurate in describing a likely outcome. It does not describe the policy of either side. Official policy is described by the word "deterrence." There are many aspects of deterrence, however. Deterrence by threat of retaliation is only one. Threat of retaliation is the ultimate deterrent to nuclear aggression. If all other deterrents fail, threat of retaliation should be the final barrier to nuclear attack from either side.

Another idea is to create uncertainty. Deterrence created by uncertainty is believed by some to be a necessary part of the barrier to nuclear conflict. I will counter several aspects of the "uncertainty" argument in subsequent sections, but the thesis is: "Creating uncertainty of a successful outcome for the attacker deters the attack." If one is thinking in traditional military fashion about win-lose outcomes, the thesis seems reasonable. I will show that it is counterproductive in most instances under modern conditions. Superior numbers and superior technology are also seen by some, particularly industrialists, as being part of deterrence. It should be rather obvious to all, however, that attempts to achieve superiority simply drive escalation. Escalation itself is a no-win game.

The goal of a policy of deterrence is to drive the probability of deliberate attack toward zero. This is a proper military objective in the name of defense. It is, in fact, one of the few military options available to "defenders" using traditional political policies. It is a time-tested idea. But as you have seen in the neighborhood example, even though the probability of deliberate attack remains near zero, the probability of accident and irrational outcomes increase with automation. There seems to be a preoccupation in the SDI arguments with enhancing deterrence, thus attempting to reduce the probability of deliberate attack. It will become obvious as we proceed that is not where the danger lies.

There are also other more constructive ways of viewing deterrence, for example: deterrence through cooperation. This policy descends from the new rules of the game.

1.5—MYTH OF THE HIGH GROUND

*Every tradition grows ever more venerable ... the reverence due
to it increases from generation to generation. The tradition finally
becomes holy and inspires awe.*

– F. W. Nietzsche, *Human, All Too Human*, 1878

If one is charging up San Juan Hill with Theodore Roosevelt in 1898,
or planting the flag on Iwo Jima in 1945, "taking the high ground" is an
appropriate battle tactic. Foot soldiers, horses, and low powered vehicles
have great difficulty battling uphill. Those on top have all the advantages.
Bullets, rockets and other non-orbiting objects come down easier when
aided by gravitation. The "high ground" in classic warfare was the best
"defendable location." However, mortars, howitzers, powerful vehicles,
personnel missiles, and tactical air support have tended to even things out,
making terrain less an obstacle in conventional war. The foot soldier still
has a problem, however.

   In nuclear war and space war, the high ground is the last place one
wants to be. The deeper one is underground and the more concrete that
exists above one's head, the safer things are. Even in air battles height is
not necessarily an advantage, even though it was until just a few years
ago. Countermeasures, such as radar and missiles, have eroded the traditional notion about the sanctity of the "high ground" Target-seeking
surface-to-air missiles directed at aircraft are much less effective against
low flying targets. And to claim that weapons in space are necessary

because that represents the "high ground" (as some presumably authoritative people have done) is completely asinine or just another propaganda ploy. Whoever originated that rationalization to justify putting weapons in space either knows nothing at all or is trying to pull the wool over someone's eyes.

There can hardly be a target more vulnerable in principle than a satellite which circles the Earth periodically on a precisely predictable orbit. Changing an orbit either up or down takes exactly the same amount of energy once the orbit is established. Going "down" is no easier than going "up." And getting into a particular orbit is equally easy (or difficult) from any part of the world; it just takes a little more or a little less fuel depending upon latitude and direction. An orbiting target in space has about the same vulnerability as if it were in a shooting gallery on Earth. No scientific breakthroughs are required, just engineering to accomplish the results. Fortunately, both East and West restrained themselves during the 1970s in developing the weapons to kill satellites, thus a reliable operational capability to do so does not yet exist (1987). In any future war in space, in the vicinity of planet Earth and with the new exotic technologies, however, humans in a space craft, or even robot space craft, would have a life expectancy of about 90 minutes, the period of one low orbit. The aircraft-like evasive maneuvers and aerobatics—the "dog fight" tactics—portrayed in science fiction movies such as "Star Wars" and "The Empire Strikes Back," are not realistic views of what spacecraft in orbits can do. They are confined rather severely to their orbits, except with large expenditures of propellant. No amount of new science is going to repeal the constraints imposed by the planet's gravitational field.

But perhaps the greatest fallacy of viewing space as the "high ground" is the confusion of believing space is a defendable location. Space is a medium for achieving mobility. No military planner should ever have made that mistake, yet it appears that this error was very fundamental in the initial belief that SDI might work: that space could serve as a defensive position. Space stations will require a peaceful environment in order to survive for very long.

Trying to control the oceans serves to illustrate the problem. A carrier task force is an example. The main strike arm, the aircraft aboard the carriers and the carriers themselves, must be protected and supported by numerous additional ships and other aircraft in order to accomplish their strike mission. Of the total cost of the task force, at least 70% is required for the support and defensive role. In other words, it takes more resources to defend and support the strike arm than it does to deploy it. It cannot survive in battle alone, therefore additional protection and support is necessary. The space problem is even more forbidding. Although exact numbers cannot yet be calculated, it would be reasonable to expect that orbiting battle stations would require 90% or more of their assets to be devoted to their own defense. Thus, for every dollar spent on building a satellite, nine additional dollars would be required to defend it. Even that amount is not likely to yield sufficient protection because of the unrelenting vulnerability of space objects to the very technology required for SDI. The initial estimates for SDI, as proposed, exceeded $500 billion, and some approach one trillion dollars for just the antimissile weapons. But the need for defending virtually indefensible space objects would multiply that requirement several times.

The myth of the high ground is just another anachronism left over from the past—the recent past admittedly, but the past, nevertheless. Applying the idea to the space environment has been a colossal error in military thinking. The environment itself simply precludes using such a notion effectively.

1.6—SATELLITES

Twinkle, twinkle little star,
How I wonder what you are
You move so quick across the sky,
Is it perhaps that you're a spy?

— Nursery rhyme, 1806

Since the earliest probes into space, indeed even with the coming of ballistic missiles, the Department of Defense has been studying the most effective way to utilize space technology to carry out its assigned mission of providing for the nation's security. The stepped-up emphasis on SDI has brought to the fore, with increased priority and funding, studies that have been progressing for years.

The most well-known and most successful programs developed during the past thirty years are those associated with surveillance and intelligence gathering, i.e., "spy" satellites. These have, during the era of space flight, begun to establish the idea that "there can be no secrets" and to establish the stabilizing effect of more certain knowledge. They force us toward candor and openness. Anything that is on the surface of the Earth can be observed and reported in significant detail daily. The types of sensors, their accuracy, their ability to discriminate and to report events on the surface have improved continuously through the relentless march of technology. New construction; movements of people and troops, cars, trucks, supplies; changes in foliage in fields, a new ditch dug for a missile emplacement, can be monitored with great precision. Even nighttime presents only a relatively small obstacle. No significant military activity can be carried out in the open anywhere on the surface of the Earth without being discovered by friend and foe alike. Even most camouflage attempts can be defeated by orbiting sensors. At this point in history only underwater, underground, and activity within buildings cannot be directly observed by a multiplicity of orbital devices. The time scale of orbits, of the order of 100 minutes, assures that changes going on below appear like time lapse photography. The details of a yearlong operation can be made to appear like a flower opening up overnight—but in seconds.

The foregoing facts about satellite technology represent the greatest advance in security against conventional warfare in all human history. The ability of the world's armies to amass on a battlefield, to move and to totally surprise the opponent is impossible. Because fear of what an aggressive adversary might be able to do undetected has always caused nightmares for military commanders, satellite surveillance has helped change the character of war. Total surprise is no longer possible unless weapons are permitted into space. Further, such information tends to reduce (but not eliminate) the compulsion to exaggerate the enemies' capabilities and under rate one's own just to be safe. The stabilizing effect of surveillance and the progress of ever more discriminating sensors represent a valid area of military activity in space. Increased information and the sharing of information through cooperative ventures represent one important facet of the new rules of the game.

But satellites are not the entire solution either. They have areas of weakness, as does all machinery. And naturally, rushing to discover countermeasures against those weaknesses began with the design of the first satellite—on both sides. The stabilizing effect of surveillance satellites was not long in being recognized, however. Even though anti-satellite weapons (ASATs) were designed and built (1958) almost as quickly as the first satellites themselves, it soon became apparent that to threaten a surveillance satellite with an anti-satellite missile could set off the very situation that needed to be avoided—massive retaliation. Furthermore, just the existence of an ASAT was (and still is) destabilizing, because neither side can distinguish accurately between a normal failure in the equipment and its loss due to enemy action. Even a destructive meteor hit on one's satellite might be construed as prelude to a nuclear attack if

ASAT weapons were known to exist and the political situation were tense. It became clear that continuing the development of ASAT weaponry was not in the best interest of either side. The Anti-Ballistic Missile Treaty of 1972 (see Appendix A3), without specifically mentioning ASAT systems, effectively precluded ASAT development, because observation satellites are considered part of the "national technical means for verification" of the provisions of the Treaty. The Treaty provisions require that neither side will interfere with the other's "national technical means for verification." Unfortunately, however, there is nothing in the Treaty language that specifically prohibits developing and testing ASAT weaponry.

It is this loophole in the Treaty that permitted the development and testing of U.S. ASAT systems against points in space in 1985, after thirteen years of almost no activity in the ASAT area by either side. This hiatus in ASAT activity was viewed with alarm by the U.S. military because their system was obsolete and ineffective. The fact that the Soviet ASATs are even older and even less effective has never been mentioned in the drive to obtain funds and approvals for new weapons. The lack of ASATs, until the current U.S. activity, was, in fact, a very stable and safe condition. To acquire the ability to threaten unarmed satellites is a very foolish notion for either side. The deployment of weapons in space, however, will require that it be done in the name of "defense," and, further, that surveillance satellites be armed in a fruitless effort to defend themselves.

1.6.1—SATELLITE VULNERABILITIES

They move through the silence of space on unseen wings, pulled only by the thread of gravitation: tenuous, expandable, relentless. The serenity and beauty belie the eternal silence that awaits the error.

— Dr. Edgar Mitchell

From the earliest days of space flight the most severe limitation has been the amount of machinery and fuel required to place a payload in orbit. Payloads representing no more than 5% of total launch weight were normal in the beginning days of space exploration. Depending upon the orbit one needs to achieve and the state of existing rocket engine technology, the payload amount can vary from a low of 1% up to approximately 25%. To achieve the larger payload fractions requires huge booster rockets and multiple stage boosting. The laws of physics and the gravitational field just won't yield in this regard. The Earth's gravitational field is like a deep well from which one must first be thrust before useful things can be done. Propelling out of the well consumes most of the energy required to get out and stay in orbit. This energy is converted into the kinetic energy of velocity.

The problem is always a trade-off between the high thrust needed to get launched and the higher efficiency of fuel usage once in orbit. The two cannot be combined. Only chemical fuel rockets or nuclear-powered rockets can produce sufficient initial thrust to begin the journey from a launch site. But nuclear fuel rockets are undesirable for many reasons, so that leaves just chemical rockets for launch, in which case most of the launch weight must be the fuel. Higher efficiencies can be obtained with lower thrust engines once orbit is attained. But however one looks at the problem, the cost of placing one pound in orbit has not been reduced below $1000 as of this writing. The early hope for the shuttle design was to reduce costs (1970 prices) to about $100 per pound. Inflation has more than offset technical improvements, however, and costs have risen to about $3000 per pound. At this rate, the launch costs alone of placing in orbit a battle station the size of a 747 aircraft would reach about $750 million.

Weight and cost, therefore, are critical factors. The spindly, flimsy structures launched into space were designed that way because once in orbit they do not feel the "weight" caused by gravitation and do not need as much structural rigidity. But as importantly, every pound of unnecessary material represents an enormous launch penalty. When we think about protecting equipment from battle damage by hardening the structure or applying thermal-resistant coatings against lasers, the weights and the costs begin to shoot up rapidly.

It may distress space buffs, but the gigantic, armor-plated juggernauts depicted in science fiction fantasies would need to be constructed or assembled in space and at unbelievable costs if indeed they became feasible at some point in the future. Every pound going up starts out as 7 to 10 pounds and large items must be taken up in small pieces. Any fuel to power the machinery suffers the same penalty unless it can come from natural solar energy or efficient nuclear sources. This is the first fact of life about space flight. Weight is exceedingly precious! Changes in technology may provide small changes in payload ratios but basically it is a problem of brute force to overcome the Earth's gravitational field. That will not change regardless of technology or new science.

Next, all equipment is sophisticated, relatively delicate, highly compacted in volume and must be protected against the temperature extremes and the vacuum of the environment. Protecting such equipment against battle damage would be next to impossible. Techniques for hardening structure against impact require additional weight and are only marginally effective against the kinetic energy of other objects in orbit. The tiniest pinhole in a pressure container is deadly. A tiny particle of metal floating into a circuit board destroys it. Electromagnetic pulses from any number of countermeasure weapons can damage circuitry with relative ease. Even the sophisticated high performance jet aircraft in use in the world today are far more vulnerable to damage from minor problems and to being operationally disabled than their predecessors of World War II. All the exotic military aircraft of the world are in reality better utilized as peacekeeping machinery against a lesser equipped opponent. They can deliver impressive firepower but are vulnerable to equally sophisticated countermeasures. Their ability to withstand battle damage and remain in action is quite restricted.

Even more is this true of space equipment. The more sophisticated and exotic the equipment, the less is the natural survivability and the more effort and resources must be devoted to assuring survivability. It takes millions of dollars of time, effort and resources to build any instrument capable of functioning in space. A primitive cannon ball projected with orbital velocity (which it must have to be there in the first place) can destroy any of them. This is the second fact of life about space vehicles and war. Regardless of the sophistication of the space equipment, building it is far more difficult and expensive than building the machinery needed to destroy it. The minimum technologies needed for destruction are detection technologies (radar), launch technology and guidance technology. All these have become quite sophisticated in this era. The lack of operational ASATs in the arsenals of East and West is not because they are infeasible but because, fortunately, they have not been a high priority development during the years of operating under the ABM Treaty.

In addition to the foregoing problem of natural survivability, any object in orbit is in a precisely predictable path. It returns to the same point in the orbit about every 90 minutes (low orbits). It can have only limited maneuvering fuel for evasion because of the weight penalty. But even if it maneuvers it must use the same or even more fuel, never less, to get back on its operational trajectory, thus evading an attack may put it completely on the other side of the world from where it needs to be to accomplish the mission. The problem of evasive action and fuel conservation is a huge problem for satellites, probably so much so that evasive action is not an appropriate tactic against relentless attack. A satellite in Earth orbit is more vulnerable to tracking and attack than anything else on, above, or below the surface. It is at least an order of magnitude easier to kill them than it is to protect them. Or if it attempts to evade the attack, it is no longer in orbit to accomplish its mission. Only satellites in exceedingly high orbits, a few thousand miles out in space, become difficult to attack. They, however, can't do much harm from those distances either. The lower and more effective they are as ABM weapons, the more they are subject to attack. It might appear that a new twist to an old profession could arise in the space age—the mercenary space warrior. By being paid a minor fraction of the costs of the satellites to be killed or diverted he could supply his own ASAT's and still be the richest man on Earth. The task of protecting a satellite will be far more difficult than designing the equipment it is sent up to carry.

There are only three variables available to counteract the vulnerability of satellites. Two of those concern detection. They are: a small area of visible profile and the material from which they are constructed. Small targets or those oriented to minimize the area illuminated by radar are hard to detect at great distances. Illuminating a target at 1000 miles and obtaining a usable return signal is directly dependent on the size, orientation and material of the target. For small targets, locating them in the vastness of space is difficult. But because a battle station cannot be too small and still carry the equipment needed to be effective, the problem is bounded. Larger radar arrays are all that is necessary. Radar technology is a mature technology with precisely known capabilities. The only new and innovative way to partially defeat it is with use of materials that present a low "radar profile." It is this type of material that is proposed for the stealth bombers. It could conceivably be successfully used in satellites. However, whatever is a feasible material to defeat radar detection of satellites is also a feasible material to defeat radar sensing of other vehicles such as reentry warheads and the space buses that carry them. Regardless of how one looks at the problem, detecting satellites from the ground is less difficult than detecting the smaller nuclear warheads either from space or from the ground.

A third way of overcoming satellite vulnerability is saturation of the defenses. I will discuss the "numbers game" in a subsequent section regarding the killing of missiles but will touch upon it here with regard to satellites. One approach in SDI thinking is that with a sufficient number of missile killers in orbit, any ASAT capability of the opponent could be saturated and the ballistic missile defenses still retain effectiveness. Indeed, as we will point out, saturation of defenses is normally the easiest and most economical technique for overcoming a defense. But in the case of trying to overwhelm ground based ASATs with superior numbers of orbiting satellites, the costs become defeating since satellites are so much more expensive to build than are the ground-based machines needed to kill them. Saturation of ASATs with superior numbers of satellites or even orbiting decoys is not a very feasible approach.

I can find no configuration for satellites whereby the problem of detecting them and destroying them is not a more simple, thus less expensive, problem than detecting and killing warheads before they fall on a target. The economics vastly favor ground-based machines to kill satellites over orbiting machines to kill surface targets or rising missiles.

CHAPTER 2—PLAYERS IN THE GAME

2.1—A SIMPLE PICTURE

U.S.–Soviet Scientific Exchanges: Government Agreements. Summary of findings by the Friends Committee on National Legislation

The people have always some champion whom they set over them and nurse into greatness ... This and no other is the root from which a tyrant springs; when he first appears he is a protector.

— Plato, *The Republic*, Bk VII

The earlier example of two warring neighbors was sufficient to illustrate quite simply and with reasonable fidelity to the international "game" how a policy of deterrence by the threat of retaliation resulted in MAD. It is not sufficient, however, to accurately portray all the significant conditions surrounding SDI that exists between the two superpowers, the United States and the Soviet Union. To improve the fidelity of our analogy it is necessary to introduce additional characters and a slightly different setting.

Assume two large groups of people are on either side of a very large room with all their worldly possessions. They each have their champion who is armed only with powerful grenades and a grenade launcher. Each has plenty of spare ammunition. The two champions are professionals. They not only are armed but are wearing full body armor so that they themselves are protected. The groups of people themselves are neither armed nor protected nor can they leave the room. (Don't ask how they all got themselves into this situation. Probably by the same process that the warring neighbors did—they didn't like the way each other behaved.)

Each group is angry and distrustful toward the other. One group are "good guys" and the other are "bad guys," but to an uninformed observer it is difficult to tell quickly which is which. Their complaints sound about the same and each thinks his group are the "good guys." The only observable difference is that one champion has a weapon that look slightly older than the weapon of the other champion and his group is wearing eastern European clothes, whereas the others are wearing the latest in Paris fashion.

The dilemma: each champion would have no regrets about killing the other champion, nor really any concern about killing some of the other group. But things aren't quite that simple. If either launches his grenades, the other will also. Since the champions have armor, they may each get hurt a little but the people on either side won't be so lucky. Most, if not all, will be killed, particularly if more than just one or two grenades from either side explode. Thus, it is a stand-off. The people and assets are at risk but not their champions. And because the champion is not really at risk, response time is not critical. Each can probably take a hit and still respond when he chooses. Of course, people are killed if this happens, and some assets destroyed. For the people, the game has a lose-lose outcome if the battle starts or a win-win outcome if does not. For them there is no win-lose outcome, i.e., the people's probability of win-lose is zero. But for the champions there can still be any of the four outcomes. Our game theory has split into two different outcomes in this situation. The outcomes for the champions and those for the people must be computed differently. The outcome probabilities are independent of each other, and each set must sum to 1.0. The peoples' outcomes are only determined by whether or not the shooting starts. The champions still can have win-lose outcomes even if the shooting starts.

These champions are professional and persistent. The job is to protect the people and their assets. Each feels frustrated that he is unable to get an edge on the other. Each decides that with a more powerful armor piercing grenade and with a new telescopic sight he could precisely hit the vulnerable spot of the other where the armor is weak. With the new equipment he can kill the other champion before he launches his grenade.

There is still quite a risk, however, that if several grenades go off at once under any circumstances most on both sides will be killed. The first shot must be a good one to kill the champion but with damage to few of the other assets and none to his own people.

The champions begin to get nervous for fear that the other might get armor piercing grenades first or new gun sights first. If either champion is wounded badly or killed by the other's weapon, he may not be able to throw his grenades and thus "protect" his constituency. (We notice that the people are standing as far away from the champions as the room permits.) Moreover, the champions are still very human and the thought of the other champion having armor-piercing weapons increases the nervousness. They don't want to die either, but particularly they don't want to be outdone by the other champion. It is part of the professional "game." The armor is vulnerable to new high powered and accurate weapons, but only with the luckiest of shots would the old grenades penetrate the armor.

In actuality now the people with the champion who first gets the new weaponry are temporarily at greater risk, even though their champion is not. The other fellow doesn't have anything to pierce armor (yet). The one without the armor piercing capability starts to sweat a little more when the armor piercing weapon begins to appear on the other side because he, his weapons, and professional reputation are now personally at risk as well as his constituency. He considers: should he take a desperation shot with everything in the arsenal aimed at the opposing champion, hoping for a lucky hit? Also, hoping that not too many of his own people get killed. Or should he wait, hoping that his own new armor piercing weapons will get there soon, before the other champion is ready to shoot?

Whichever he does, he is wrong. If he shoots first, he is the "bad guy" but maybe still alive with some of his people and with reputation preserved. If he doesn't shoot first or if he shoots and misses, he is the "dead guy"—and so are lots of his people. And if he throws too many, all will be dead except perhaps the champion with the new equipment. Thus it is clear that the introduction of armor-piercing weaponry elevates the tension and the likelihood of a preemptive strike by the opposing side.

In the vernacular of game theory terminology, the champion who first got the new weapon improved his own probability for a win outcome, but the probability of lose-lose increased for the people. Only when both had new weapons did the win-win probability for the champions become equal again. But the champions' probability of lose-lose increased over their win-win probability because they were now personally at risk. The people's lose-lose probability is greater with the new weapons than with the old, but less than when only one has new weapons. And the lose probability is greatest for the people whose champion alone has the new weapon.

Except for one additional complication, *the foregoing situation is precisely the current nuclear situation*. The champion is military leadership. The armor represents hardened missile silos. The grenade and launcher represents older missiles of both sides which do not have sufficient guidance accuracy to be assured of killing the opposing missiles in their silos. The armor piercing grenade and new sights represents the newer guidance technology which can accurately hit the silos (or anything else). It is called attaining "first strike capability."

Because it makes the opposing champion more nervous, thus more likely to make a desperate move, first strike capability is destabilizing. The likelihood of a desperation preemptive strike is greater when only one side has it. And a preemptive strike is particularly great if the side that does not have it is aggressive or exceptionally fearful of the other. It puts the nation's population with first strike capability at higher risk although it makes that side's champion feel safer and happier. If one champion is to have it, both should in order to minimize population risk. The risk to both populations is highest when only one side has first strike capability (or any new threatening technology that must be countered). It reduces somewhat if both sides have first strike capability and parity is achieved but is still greater than if neither side had first strike capability (or new technology). These strange situations arise from playing by the old rules of the game but with modern technology, short warning times, and the risk factors of the champions becoming less than the population they are protecting. Should either champion after gaining first strike capability decide to use it, then that is a deliberate act of aggression against not only the opposing champion but the population of the other side.

In his book, *The Real Peace* (1983),[13] former president Richard Nixon states (pg. 22–23) that the Soviet land-based missiles have already achieved a first strike capability and that the U.S. does not have and has no plans to obtain one. That is unfortunately an untrue statement. The openly published material on the capability of both sides shows this statement to be demonstrably false. Improved guidance technology has always been on the side of the U.S., as has the invulnerability of the nuclear submarine arsenal. One would guess that the contemporary "champions" may have been a bit misleading even to a former president who relies upon reports rather than digging out the data himself. Both the MX missile and the Trident II represent first strike weapons. The intricacies of first strike are examined in depth in section 4.3.

The foregoing metaphoric scenario is quite accurate with regard to U.S.-Soviet dynamics in nuclear technology except that there are three champions on each side, not just one. And each has slightly different capabilities and vulnerabilities. One champion on each side has ground-based missiles in silos. The second has submarines which are almost invulnerable underwater. And the third has airplanes which can carry cruise missiles and are mobile. (Formerly they carried nuclear bombs directly.)

It is the champion armed with submarines on each side, however, which mostly preserves the peace in the late 1980s through deterrence by threat of retaliation. The submarine champion's armor (deep water), coupled with mobility, make him invulnerable. In the mid 1980s, the U.S. capability was about 60% submarine based and the Soviet less than 10%. The ability to locate and target submerged submarines is still not sufficiently precise to significantly threaten their destruction. That may not be true in another twenty years, however. Until that time submarine retaliatory capability provides stability even though first strike accuracy against targets is being developed by both sides.

But the major lesson is: With each new technological improvement, the champion who possesses it is more comfortable and happier, the other more nervous until he also has parity. But the risk to both populations keeps rising with each new technological improvement. The population risk becomes greatest when one set of champions has better technology than the opponents. And the risk is greatest to a population when their champion begins to introduce the new technology. It is the population's "window of vulnerability." The impact of improved technology is popularly thought to produce more security for the one possessing it, but that isn't the case in the nuclear age. The awesomeness of the killing power on both sides makes the state of mind—the intention and rationality of each side—paramount. But, even more, it is the ability to assess accurately the state of mind of the other side that determines outcomes. The champion's paranoia increases along with the population risk as a result of improvements in the opponent's technology. And what happens if the new technology results in reduced warning time and greater automation? The tension and the attendant risk must dramatically increase as the warning time decreases. We will examine this precisely in section 4.4.

The modernization of the weapons of both superpowers to achieve first strike capability is underway. That will not likely stop. Fortunately, the current invulnerability of submarines causes the situation to retain a certain stability. It is still not possible for either side to overwhelm the other with assurance that retaliation would not occur. But since population security now depends so much on the thinking, intentions, and reaction time of the opposing champions as well as their technology; moreover, since the priority of the champions is not necessarily the priority of the population, the population must have a lively interest in the details of any changes in the situation.

The foregoing analogies correctly embody the fundamental issues with which civilization is confronted in the nuclear missile era. They seem accurate and valid. Why then are we proceeding in the direction of SDI? The answer is not difficult: history and tradition! The champion's problem is to best protect his constituency. Champions are accustomed to thinking in terms of win-lose outcomes where if the champion wins, the people also win. They are not accustomed to thinking in terms of the people's interest being quite different. Unfortunately, the people mostly have been thinking in the same traditional fashion. Throughout all history, military wisdom has dictated acquisition of the newest and best technology to carry out its function. It seems to have worked in the past. The wisdom has been tested by thousands of years of precedent. It is not easy to put that aside.

But never before has the new medium, space, been so radically different than the conventional media of mobility, water and air. And never before has warning time been able to approach zero. Finally, never before have two nations had the capability to totally destroy each other and take a large portion of the world with them. Never before has total annihilation of two nations been within forty minutes of some irreversible, foolish mistake or accident. The situation requires an examination much more deeply and thoroughly than any confrontation in the history of humankind.

Let's list the major questions explicitly which should be carefully considered in assessing the actions of the "champions" as they confront each other and as they call for newer technology. The questions were derived from a detailed study of the actual confrontational conditions. The foregoing analogies were created to illustrate the need for the questions.

◊ What are the real motivations of the leadership of both sides: protecting population or retaliation capability?

◊ What is the decision-making machinery that determines whether or not to shoot?

◊ What are the warning times that a technology creates for the decision makers?

◊ What is the retaliation decision time if different from a warning time?

◊ Is a new technology stabilizing or does it put the population (and thus the world) more at risk?

◊ Can a new technology deliver what it promises?

◊ Does a period exist wherein a desperation preemptive strike is likely to take place?

◊ Although not embedded in the simple scenarios, what role is the economics of armament manufacturing playing in these decisions? Could this factor be driving the system?

◊ Are there any plans to use nuclear weapons to attempt to achieve limited military objectives or in support of national policy?

◊ Is the real threat aggressive intent or is it accidental nuclear war?

Technological benefits and economic trade-offs should not be attempted until the answers to the foregoing questions are well understood. They are the key issues.

The unprecedented "pre-sell" emphasis given to the Strategic Defense Initiative suggests that we should examine its potential in light of the foregoing questions. Subsequent sections will examine the consequences of the technologies with the above questions in mind. But first let's construct an additional simple scenario to obtain a better grasp of the situation.

Scenario:

Consider that one of the champions (the one defending the crowd dressed in the latest Paris fashions) believes that he can make a rifle that is powerful enough, accurate enough, and fast enough to shoot any bullets or grenades of the opposing champion before they can fly across the room and explode. His bullets will not harm the opposing champion or the opposing group, so he says. It will just kill bullets and grenades if they are launched across the room. The champion announces his plan to both sides and sets out to build such a weapon. The opposing champion (the one defending the crowd in eastern European clothing) reacts in a way suggesting great alarm at this idea.

What do we now do? Back our champion or oppose his idea? Is such a weapon possible? If it is possible, is his aim and reaction time quick enough to use the weapon? Is the opposing champion likely to become too desperate in the process and launch a preemptive strike?

Let's consider the essential features of the actual situation. The trajectory of silo-based ICBMs between the Soviet Union and the United States requires less than 35 minutes to complete. Depending upon the particular missile characteristics, approximately 5 minutes or less is needed for launch, 25 minutes for coasting in a sub-orbital trajectory, a couple of minutes to reenter and strike the target.

With infrared wavelength sensors, missiles can be automatically detected during their launch phase from great distances out in space. The hot launch gases make easy identification. Then with multiple sensors to provide redundancy, detection seems reliably assured. The warning system gives approximately 30 minutes notice of an attack. But there is really nothing that can be done once a launch is detected except hide, wait, and retaliate at the appropriate time.

But our champion doesn't want to wait and waste that 30 minutes between detection and impact. He wants to kill those incoming nuclear weapons before they explode. And the more accurate their guidance systems which threaten the ability to retaliate, the more eager the champion is to kill them. Let's look at what is involved in killing the incoming missiles.

Each missile is capable of carrying up to 10 reentry vehicles (RV) with nuclear warheads. It is also capable of carrying at least 10 decoys per RV. Thus, one missile launched blossoms into at least 110 targets in space, 10 of which need to be distinguished from the other 100 and destroyed, if possible, before exploding on their targets. Ten accurately guided warheads can with high probability destroy 5 missile sites (two explosions per site) or 10 cities. But the fallout from 10 nuclear bombs will certainly contaminate most of the country even if targeted on remote missile silos.

Certainly then, the champion's problem of protecting the retaliation capability and the population is most easily solved if he can reliably kill the one missile rising from its silo over the aggressor's territory rather than wait to distinguish 10 RVs from 100 decoys and be required to kill the 10 RV's as they enter home territory. IF THIS IS POSSIBLE TO DO, it seems logical, appropriate, and should by all means be done.

Most likely, it was a similarly over-simplistic picture, such as the accompanying sketch on a napkin that the French Minister of War used to get approval to begin planning the Maginot Line in WWII. The line has since become a metaphor for expensive efforts that offer a false sense of security.

Throughout history, major programs of governments, militaries, and most large organizations have hinged on obtaining high level approvals in such rapid and simplistic ways. One draws the conclusion after sorting through the conflicting claims and counterclaims about the SDI program that it also began in a similar fashion.

In the following sections we will examine some additional subtle but profound difficulties underlying this hasty and erroneous decision.

2.2—THE PRESIDENT'S DREAM

Vain the ambitions of kings
Who seek by trophies and dead things
To leave a living name behind,
And weave but nets to catch the wind.

— John Webster, *The Devil's Law Case*, 1623

While doing research for this chapter, I was reminded of a story told by a member of the British royal entourage. When a member of the royal family goes abroad on an official state visit it is the task of the staff to make certain that the royal person appears in public as a "stately white swan gliding serenely across the pond of public events." "But," said the member of the staff, "the public never knows all the furious paddling that is going on beneath the surface."

It is likewise true for any head of state that a calm and confident public posture heralds an enormous amount of "dog" work going on elsewhere. However the pond of public events is not always serene. The rocks and shoals of public life, together with the beasts of prey that enjoy swans, make public life difficult. The stateliness of the swan can be ruffled if the paddling down below gets confused and lacks coordination. And sometimes the swan is steered in a totally erroneous direction. Then the confusion increases.

Such is the case for President Reagan and his Strategic Defense

Initiative. The lack of in-depth coordination down below began to emerge from the outset. It became increasingly apparent as technology professionals studied the defense issue in depth that serious problems existed. *The New York Times* consistently ran articles over many months opposing this initiative with the tempo of criticism increasing in the winter and spring of 1987. Newsweek published a significant dissent as well in early spring 1987. After November 1986 the controversy around and investigation of the Iran-Contra crisis, dominated most of the nation's media coverage for many months. Surveys of public opinion in the winter-spring of 1987 revealed that a significant majority (60%) of the public still favored continuation of the SDI. The disparity between professional and public viewpoints attested to the President's personal charisma and to the fact that the public had not been adequately informed of the dangers nor of the alternatives by early spring 1987.

With this in mind, it seems instructive to review what has been the presidential position.

On March 23, 1983, the President Reagan announced his vision for a move toward a defensive strategy which would assure freedom from attack by nuclear tipped missiles and would serve subsequently as a basis for movement toward total arms reduction. That announcement is sufficiently important to quote in full that portion of the text dealing with this issue in order to remind you how this began. Much of the population still seems to cling to these comforting words even though the actual situation is vastly different. Certain portions of the text are underlined here for subsequent comment.

"... Now, thus far tonight I have shared with you my thoughts on the problems of national security we must face together. My predecessors in the Oval Office have appeared before you on other occasions to describe the threat posed by Soviet power and have proposed steps to address that threat. But since the advent of nuclear weapons, those steps have been directed toward deterrence of aggression through the promise of retaliation [the notion that no rational nation would launch an attack that would inevitably result in unacceptable losses to themselves].

"This approach to stability through offensive threat has worked. We and our allies have succeeded in preventing nuclear war for three decades. In recent months, however, my advisors, including in particular the Joint Chiefs of Staff, have underscored the necessity to break out of a future that relies solely on offensive retaliation for our security.

"Over the course of these discussions, I have become more and more deeply convinced that the human spirit must be capable of rising above dealing with other nations and human beings by threatening their existence. Feeling this way, I believe we must thoroughly examine every opportunity for reducing tensions and for introducing greater stability into the strategic calculus on both sides.

"One of the most important contributions we can make is, of course, to lower the level of all arms, and particularly nuclear arms. We're engaged right now in several negotiations with the Soviet Union to bring about a mutual reduction of weapons. I will report to you a week from tomorrow my thoughts on that score. But let me just say I am totally committed to this course.

"If the Soviet Union will join us in our effort to achieve major arms reduction, we will have succeeded in stabilizing the nuclear balance. Nevertheless, it will still be necessary to rely on the specter of retaliation, on mutual threat. And that's a sad commentary on the human condition. Wouldn't it be better to save lives than to avenge them? Are we not capable of demonstrating our peaceful intentions by applying all our abilities and our ingenuity to achieving a truly lasting stability? I think we are. Indeed, we must.

"After careful consultation with my advisors, including the Joint Chiefs of Staff, I believe there is a way. Let me share with you a vision of the future which offers hope. It is that we embark on a program to counter the awesome Soviet missile threat with measures that are defensive. Let us turn to the very strengths in technology that spawned our great industrial base and that have given us the quality of life we enjoy today.

"[Up until now we have increasingly based our strategy of deterrence upon the threat of retaliation.] What if free people could live secure in the knowledge that their security did not rest upon the threat of instant U.S. retaliation to deter a Soviet attack; that we could intercept and destroy strategic ballistic missiles before they reached our own soil or that of our allies? "I know this is a formidable, technical task, one that may not be accomplished before the end of this century. Yet, current technology has attained a level of sophistication where it's reasonable for us to begin this effort. It will take years, probably decades, of effort on many fronts. There will be failures and setbacks, just as there will be successes and breakthroughs. And as we proceed, we must remain constant in preserving the nuclear deterrent and maintaining a solid capability for flexible response. But is it not worth every investment necessary to free the world from the threat of nuclear war? We know it is.

"In the meantime, we will continue to pursue real reductions in nuclear arms, negotiating from a position of strength that can be ensured only by modernizing our strategic forces. At the same time, we must take steps to reduce the risk of a conventional military conflict escalating to nuclear war by improving our nonnuclear capabilities.

"America does possess—now—the technologies to attain very significant improvements in the effectiveness of our conventional, nonnuclear forces. Proceeding boldly with these new technologies, we can significantly reduce any incentive that the Soviet Union may have to threaten attack against the United States or its allies.

"As we pursue our goal of defensive technologies, we recognize that our allies rely upon our strategic offensive power to deter attacks against them. Their vital interests and ours are inextricably linked. Their safety and ours are one. And no change in technology can or will alter that reality. We must and shall continue to honor our commitments.

"I clearly recognize that defensive systems have limitations and raise certain problems and ambiguities. If paired with offensive systems, they can be viewed as fostering an aggressive policy, and no one wants that. But with these considerations firmly in mind, I call upon the scientific community in our country, those who gave us nuclear weapons, to turn their great talents to the cause of mankind and world peace, to give us the means of rendering these nuclear weapons impotent and obsolete.

"Tonight, <u>consistent with our obligations under the ABM Treaty</u> and recognizing the need for closer consultation with our allies, I'm taking an important first step. I am directing a comprehensive and intensive effort to define a long-term research and development program to begin to achieve our ultimate goal of eliminating the threat posed by strategic nuclear missiles. This could pave the way for arms control measures to eliminate the weapons themselves. <u>We seek neither military superiority nor political advantage.</u> Our only purpose—one all people share—is to search for ways to reduce the danger of nuclear war.

"My fellow Americans, tonight we are launching <u>an effort which holds the promise of changing the course of human history.</u> There will be risks, and results take time. But I believe we can do it. As we cross this threshold, I ask for your prayers and your support.

"Thank you, good night, and God bless you."

From the way we humans are accustomed to thinking about war, and particularly the way military people are accustomed to thinking about war, the foregoing address is magnificent. It could hardly appeal to the American public in a more compelling way. It addresses the patriotic pride, the true public desire for peace, the pride of our technological and industrial prowess and brings inspiration for a hopeful future. If words alone could accomplish the goals they seek to inspire, if modern technical reality were indeed so simple, the dream could possibly have been realized. If the word "nuclear" is removed and the speech transplanted to early in this century it would have been a wonderfully progressive concept well ahead of its time rather than hopelessly behind the technical realities. It is what some national leader should have proposed before the nuclear age began, not after the space age is already well advanced and Pandora's Box opened. The technological reality of the twentieth century completely changes the way in which the dream of a less hostile world can be accomplished. The battlefield of space does not admit the same rules of the game that could be used pre-1945. One cannot think in terms of any prior wars when considering what "defense" or "wars in space" will demand. The old rules do not apply, but it would not seem that senior leaders on either side have fully grasped the significance of this fact. The rhetoric remains much as in the past and the moves and countermoves remain unchanged. But the environment causes all that to be obsolete.

President Reagan amplified his thinking on several occasions, that is, of wanting to find an impenetrable shield:

" … So, we were all in agreement that it was worth us starting out to find if we could find a weapon that could intercept those missiles, and intercept them thoroughly enough, not just like having anti-aircraft guns. Some of the bombers always get through. No, to really stop them."

– U.S., President, Interview with Hearst Corporation Editors, 30 October 1984, *Weekly Compilation of Presidential Documents*, Vol. 20, No.44, 5 November 1984, p. 1699

"I want a defense that simply says that, if somebody starts pushing the button on those weapons, we've got a good chance of keeping all or at least the bulk of them from getting to the target … I don't think of it in terms of let's put it around this place or that place. Let's put it in such a way that those missiles aren't going to get to their target."

– David Hoffmann, "Reagan Says 'Star Wars' Effort Would Continue Despite Pact," *The Washington Post*, 13 February 1985, p. A1 8

Secretary of Defense, Casper Weinberger, reiterated President Reagan's wish three and a half years after the original announcement:

"Our goal must remain as the President first defined it: to save lives, not avenge them. SDI seeks a safety shield for all our people—a shield to deter attack, and to protect us should deterrence fail."

– "Weinberger Cites Defense Critics Flawed Reasoning," *Tech Trends International*, 8 September 1986, p. 1

From the time of the president's speech in 1983, the debate continuously raged over the possibilities and impossibilities of the proposal.

The picture of what could realistically be done with technology began to emerge from studies. The alarming development is that the drive toward a seemingly unattainable goal did not abate but increased. Moreover, the studies demonstrated that to pursue the goal makes the situation even more dangerous.

It is instructive to look at many of the statements of U.S. leadership since that initiating announcement. While the President and the Secretary of Defense continued to assure the public that the purpose of SDI is to "shield" the country (population and cities) from a rain of nuclear missiles, the persons responsible for executing the administration's policy have always talked in completely different terms. Lt. Gen. James Abrahamson (USAF), Director SDI Office, for example said:

> "I don't believe there is any such thing as a perfect weapon … finding ways to provide a better basis for deterring aggression, strengthening stability and increasing the security of the United States and its allies. Thus, the promise of enhanced deterrence is an elemental factor underlying the SDI."
>
> – *Defense Daily*, 10 December 1985, p. 195

And Richard N. Perle, Assistant Secretary of Defense for International Security Policy until the spring of 1987:

> "Assistant Defense Secretary, Richard Perle, has declared that the initial goal is 'the defense of America's capacity to retaliate.' In the lexicon of nuclear strategists that's shorthand for the defense of missiles, not cities."
>
> – "Star Wars under the gun," *U.S. News & World Report*, 21 May 1986

Numerous other similar quotes and references underscore the differences in approach between those setting policy and who publicly announced the objective, and those executing policy who must deal with technical and practical reality. It has been increasingly clear to all who studied the technical feasibilities, particularly those charged with making the dream come true, that total (population) defense is not even remotely a technical possibility, then, now, and likely never.

In addition, however, to the technical discussions, there has continuously been an aggressive rhetoric underlying announced policy that should give all cause for alarm. It has been part of the staff thinking that shapes policy announcements (the paddling going on down below). Whereas senior officials declare U.S. "defensive intentions" and peaceful policies, there are individuals who speak quite differently from what is publicly said at the top. These statements cause not only the opponents to distrust U.S. intentions, but should give pause to all persons who believe escalation of weaponry is not the way to defense much less peace:

For example, Richard N. Perle, one of those most hostile in setting policy with regard to the Soviet said:

"The sense that we and the Russians could compose our differences, reduce them to treaty constraints, enter into agreements, treaties, reflecting a set of constraints and then rely on compliance to produce a safer world—I don't agree with any of that.

– Fred Hiatt, "Perle's Distrust Shapes U.S. Policy,"
The Washington Post, 2 January 1985, p. A1

But in the closing days of April, 1987, Mr. Perle was interviewed on CNN television, having taken a slight step backward, arguing for very tough conditions on any short and middle-range nuclear weapons agreements for Europe. Enlightenment or just politics?

For another example, Edward C. Aldridge Jr., Secretary of the Air Force:

"We do not have to stretch our imagination very far to see that the nation that controls space may control the world."

– Fred Hiatt, "U.S. Prepares for the Day When War May Be Waged in Space," *The Washington Post*, 18 September 1983, p. A1

And Air Force General Robert T. Marsh:

"We should move into war-fighting capabilities … That is, ground-to-space war-fighting capabilities, space-to-space, space-to-ground."

– Ibid., P.M.

And also Colin Gray, member of advisory board for the Arms Control and Disarmament Agency (Reagan administration):

" … the United States should plan to defeat the Soviet Union and to do so at a cost that would not prohibit U.S. recovery. Washington should identify war aims that in the last resort would contemplate the destruction of Soviet political authority and the emergence of a postwar world order compatible with Western values."

– Colin Gray and Keith Payne, "Victory Is Possible,"
Foreign Policy Journal, Summer–Fall 1980

Only individuals who cannot view the possibility of a stable world with both U.S. and Soviet presence in it can talk in such terms. It is the static, tradition-bound world of the past. But that world is one in which the escalation of military technology and the "numbers game" using that technology proceeds endlessly toward the battle of Armageddon. (There are some in and around the periphery of the halls of power who privately avow that such a battle is their goal.) The probability of error, miscalculation, and accidental nuclear war increase correspondingly at each level of escalation of measure and countermeasure when such thinking is abroad. It projects a world that can only teeter precariously on the brink of global annihilation. It is the world view of hostile, angry men who only view peace as a possibility after conquest and domination on their own terms. These are the persons carrying out the "defensive" programs for the U.S. There seems to be no comprehension that for the people and national assets a win-lose outcome is not possible. The respective champions may think in terms of win-lose but it has no meaning in any larger sense.

2.3—THE RESPONSES

Diversity of opinion within the framework of loyalty to our free society is not only basic to a university but to the entire nation.

— James B. Conant, *Education in a Divided World*, 1948

Numbers of officials from former administrations, both right and left, have joined in the debates on the validity of the SDI concept. Their remarks are worth noting.

James Schlesinger, former Secretary of Defense:

"And let me go deal with the promise that was held out in the President's words: there is no hope. There is no hope. No realistic hope that we will be able ever again to protect American cities. There is no leak-proof defense."

— Address to National Security Issues Symposium, 25 October 1984

Henry A. Kissinger, former Secretary of State and National Security Advisor (Ford, Nixon):

"There is no serious likelihood of removing the nuclear threat from our cities in our lifetime or in the lifetime of our children."

— *Science*, 9 November 1984, p. 673

Henry A. Kissinger:

"Even granting—as I do—that a perfect defense of our population is almost certainly unattainable, the existence of some defense means that the attacker must plan on saturating it." [Kissinger was at the time pro-SDI.]

> – "Should We Try to Defend Against Russia's Missiles"
> *The Washington Post*, 23 September 1984

Walter F. Mondale, former U.S. vice-president (Carter administration):

"When Mr. Reagan explains Star Wars, it is as comforting as listening to a bedtime story. Once upon a time there was an evil empire that threatened us with terrible weapons. But then one day our side discovered a magical invisible shield. When we stretched it across our country no missiles could penetrate it. From that day on we stopped worrying about nuclear war and lived happily ever after.

"I can think of only one reason to support Star Wars. Fairy tales are often more appealing than reality."

> – Bernard Weinraub, "Mondale Asks Bank on Arms in Space,"
> *The New York Times*, 25 April 1984, p. 1

Robert S. McNamara, Defense Secretary under Kennedy; McGeorge Bundy, National Security Advisor under Kennedy and Johnson; George F. Kennan, Ambassador to the Soviet Union under Kennedy; Gerard Smith, chief U.S. negotiator in Strategic Arms Limitation Talks (SALT I) talks under Nixon:

"[SDI] offers no prospect for a leak-proof defense against strategic ballistic missiles alone, and it entirely excludes ... any effort to limit the effectiveness of other systems—bomber, aircraft, cruise missiles.

"We believe the president's so-called Star Wars Initiative to be a classic case of good intentions that will have bad results because they do not respect reality ... a solution that combines surface plausibility and intrinsic absurdity ... a complete misreading of the relation between threat and response in the nuclear decisions of the superpowers … (That is, the Star Wars plan will) "allow us to defend some of our weapons and other military assets, and so, somehow, restrain the arms race … (this) is equally unattainable (and would) destroy the Anti-Ballistic Missile (ABM) Treaty, our most important arms control agreement, (and) will directly stimulate both offensive and defensive systems on the Soviet side."

– Murrey Marder "Reagan Must Pick 'Star Wars' or Arms Talks Ex-Officials Say," *The Washington Post*, 27 November 1984, p. A5

Harold Brown, former defense secretary to Nixon:

"Technology does not offer even a reasonable prospect of a successful population defense." [The purpose of the Star Wars plan would be to] "ensure the U.S. strategic retaliatory capacity" [this means] "not defending people but defending missiles which threaten to destroy people." [Reagan and other political and military leaders] "should publicly acknowledge that there is no realistic prospect of a successful population defense, certainly for many decades and probably never."

Herbert Scoville Jr., former intelligence and arms control official:

"It [SDI] will undoubtedly set off an arms race of a kind we've never seen before."

– *Christian Science Monitor*, 29 October 1984

From analysis of the existing numbers game and the proposed technologies, it would seem that only the President and the Secretary of Defense maintain the rhetoric that anything more than missile sites' retaliation potential can be protected by SDI. There seems to be no informed opinion that has ever thought that population or cities can be protected. There seems, further, to be controversy as to whether deterrence is enhanced by SDI or totally destabilized. In reality, the answer to the last point can only be found in the responses of the opponent: the Soviet Union. We can reason that it seems destabilizing, but the litmus test is their behavior. Aside from public rhetoric, the indications are they believe it destabilizing—as would we if the positions were reversed. The real question concerns what they will do about it.

It seems the height of folly to force an equally armed and fearful human opponent into a position which gives little alternative but to shoot first in order to "defend" himself. The analysis of the costs of countering SDI versus implementing their version of SDI should give little doubt as to how they will proceed.

The foregoing excerpts represent the thinking of those most experienced in dealing with arms confrontation during the past few decades. Regardless of their position on other aspects of SDI, they are unanimous in assessing that it cannot be a system that will protect the population or cities and that it likely will result in a new round of escalation.

There are others of influence as well who totally disagree with the existing approach. As of late winter 1987, 6500 scientists whose efforts could go toward SDI have refused to accept any research moneys associated with the program and have made public declarations of non-support. A summary document of their opinions is given in the following:

"By far the most dramatic example of misplaced R & D [research and development] resources is the Strategic Defense Initiative … the SDI has grown far beyond the program of exploratory research that prudence requires and that in fact, was going on before the SDI was announced. The chaotic and inefficient expansion of the SDI—careening toward a deployment decision in the early 1990s—is draining large quantities of money and talent from military and civilian R & D efforts of far greater merit, inevitably with adverse consequences for the deprived sectors and for the country as a whole. Not least among these consequences may be the narrowing and stifling effects of a greatly expanded presence of applied military research on our university campuses.

"The danger of nuclear war is the preeminent problem of our time. If the SDI really offered any hope of escaping from that danger or significantly reducing it, one could cheerfully accept the SDI's side effects on the universities and on the balance of productivity of American efforts in science and technology more generally, as the necessary consequences of sensible priorities— just as the societal costs of increased military spending of other kinds are accepted when real threats to the nation manifestly require it.

"But the SDI does not offer any hope of reducing the nuclear danger. The President's stated goal of protecting the population of the United States and its allies from nuclear attack is unattainable at any cost; his vision of an escape from reliance on nuclear deterrence, as well motivated as it is, is an illusion. So is the apparent conviction of many around him that security in the nuclear age is still to be found in the ever more expensive pursuit of military forces superior to those of one's adversary. Both of these beliefs are manifestations of an abiding but unfounded faith in technological/military fixes for political problems, compounded by an inexhaustible capacity to commit the fallacy of the last move—to believe that improvements in weaponry on our side will somehow fail to simulate a compensating response on the other side.

"As such, the SDI and the larger military buildup in which it is embedded are prescriptions not for safety but for continuing and expanding the arms competition in ever more dangerous directions. That is more than reason enough to oppose them. The additional damage that is being done in pursuit of these costly programs—damage to the universities, to the vitality and productivity of American research and development, to the prospects for a robust and efficient economy—is simply icing on the cake."

– *Journal of the Federation of American Scientists*, September, 1986

If 6500 scientists refuse to work on the project for reasons of conscience and belief, their reasoning should not be as lightly ignored as it has been to date.

One immediately comes to the conclusion that the furious paddling beneath the surface, to continue the previous swan analogy, goes in many different directions. It further would appear that there always has been great doubt by knowledgeable persons whether the project in any of its announced versions could meet any of its objectives in principle; and even if it is possible in principle whether it should be done at all. Is it just a bedtime story as suggested by former Vice President Mondale? Is it in part realizable in principle but a colossal stupidity to pursue further, now that the facts are in? Does it even in principle provide additional security and freedom from war? Or does it, in fact, precipitate the very event which the dream promises to prevent?

The ongoing analyses by those not caught up in the whirl of the SDI funding effort itself keep pointing ever more directly at the later conclusion: that we are moving inexorably toward provoking the exact event that the people of both sides need their champions to prevent. This latter conclusion should be quite obvious as we examine the eventual outcome of escalation with the new technologies in subsequent sections. It seems only the personal charisma of the President himself that keeps a hopeless dream alive.

CHAPTER 3—NEW TOOLS FOR THE GAME

3.1—THE NEW TECHNOLOGIES

Had Nature destined us to fly, it would provide wings, and to assault each other, the weapons. It did neither; we discovered how ourselves, and must therefore manage the consequences of our discoveries.

— Dr. Edgar Mitchell

Because the Strategic Defense Initiative has been discussed in terms of a "shield" or an "umbrella" to protect against a rain of nuclear-laden missiles, it is important to understand how that shield was originally conceived to operate.

3.1.1—THE NUMBERS GAME

It may be true that people who are merely mathematicians have certain specific shortcomings; however, that is not the fault of mathematics, but is true of every exclusive occupation.

— Carl Gauss, Letter to H.C. Schumacher, 1845

Analyses have shown it must be described as a "leaky umbrella" at best because one is dealing with such large numbers of nuclear warheads, in excess of 10,000 on each side in 1987 (includes ICBM, submarine and bomber warheads). This number will continuously increase during the next decade, more than doubling if SALT constraints are abandoned, growing by about 30% if the constraints are observed. With such large numbers, planning the defense against them must be based largely upon the mathematics of probability and statistics for each of the delivery modes. The alternative to a statistical approach would be one of discrete accounting for every warhead that must be countered. Under any scheme of discrete accounting, the existing location must be continuously known for each warhead from some base point forward in time. (It is called "birth to death" tracking). And the opponent is not likely to cooperate in that matter. Therefore the first knowledge a defensive system can have of the problem is when targets begin to appear on its detectors. At that point discrete accounting should begin for each target to be engaged. The system must be designed, however, to engage the entire opposing arsenal simultaneously. And as the opposing arsenal grows, the counter system must likewise grow. The size of that arsenal can only be known approximately and within a certain probability of error. We are forced, therefore, into reliance on statistical methods.

Statistical umbrellas are always leaky. And the most immediately obvious countermeasure for a defense which of necessity is based upon a statistical approach is to increase the number of targets it must acquire, track, and destroy. With the existing arsenals, we discover that a defensive system must be 99.99% effective in order to only allow one nuclear warhead out of 10,000 to reach its target. There are currently silo based (ICBM), submarine based (SLBM), and aircraft launched delivery modes. SDI only concerns itself with the first two methods of delivery which utilize space as the flight medium. The Soviet Union possessed approximately 9000 of these warheads in 1985.[14] The aircraft carrier and cruise missile delivery modes are not addressed by SDI but currently account for less than 1000 warheads in the Soviet arsenal. The opponent would, of course, increase the percentage of these if they were not strongly countered. These likewise must be dealt with in any defensive scheme and constitute one original weakness of the SDI approach. However, to overcome this weakness a little-known Air Defense Initiative (ADI) has begun. It deals with cruise missiles and ballistic missiles with flattened trajectories which remain in the atmosphere. To the extent that space-based technologies are used for ADI, the same arguments that we use against SDI will apply.

If one desires to prevent even one nuclear warhead from falling on the population as was initially proposed, i.e., 100% effectiveness, then reliabilities and design criteria must be considerably better than has ever before been achieved. Even with massive over-design to make the kill probability approach 100%, the opponent only must throw additional warheads (or decoys) in order to defeat the system. But reliabilities approaching 100% (i.e., perfection) are simply not attainable in any area of technological endeavor. Even reliabilities above 95% are a staggering thought for a system that is based on untried technology, which must remain in space for long periods of time unattended and which can only be tested completely the first time it is used during a war of annihilation. Reliabilities and kill ratios are more properly estimated in the 70% range for each phase of the problem and are most unlikely to exceed 95% cumulatively for all phases in an actual war, even if one were to design for 99%. Various authorities in their calculations come to even lower percentages, suggesting a 90% kill rate as a realistic maximum against existing nuclear missile alignments.

In actuality, what reliability would finally result is hardly more than an educated guess on anyone's part, but it certainly will not approach 100% even with the most meticulous design and fabrication.

Thus from the outset, considering the technological breakthroughs required for a "leak proof" umbrella to protect against nuclear missiles, one is staggered by the audacity and naiveté of the original suggestion to protect the population by any such defensive method. Even 10 impacts on 10 major cities would likely destroy upwards of 10 million people outright plus the fallout and aftermath damage. This requires an effectiveness of 99.9% against existing arsenals just to limit population damage to this amount. (So far, the threat of retaliation has limited damage to zero.) If one foregoes protecting the population, quietly backs away from the original promises and instead concentrates on protecting the retaliatory capability of the military, then realistic numbers for technological reliability can be obtained. After four years of study, it is clear to everyone that has computed the problem that reliabilities sufficient to protect the population are not realistic. All authorities except the President himself had, by 1987, abandoned any talk of population defense. Unfortunately, the people are still believing that myth and relying upon those words.

Under the conditions of protecting retaliatory capability, numbers of 95% for total effectiveness have been used. Even proponents of SDI do not now use numbers greater than this in their arguments and some even use less. These are not unattainable numbers in engineering practice for mature systems, but even 95% stretches one's imagination for a new and massive system like SDI. This number could be interpreted to mean that the retaliatory system has a capability of withstanding 450 nuclear warheads and still deliver massive retaliation. Or the effectiveness number could also mean that 95% effective is the best one can realistically expect from any system such as SDI. In effectiveness against 11,000 nuclear warheads, a much larger proportion of the 1000 added warheads will get through as saturation of the system occurs. There is no way to play the numbers game except by an escalation in the numbers of measure or countermeasure. But what happens also is that as the numbers increase, the effectiveness and reliability numbers required begin to approach unattainable percentages. The reason for this is that the land, the defenses, and the population can only withstand a small number of "hits" from nuclear weapons before all life and life sustaining capability is destroyed for indeterminate times into the future. Consider the alarming effect on western Europe from the leak (not a nuclear explosion just a normal explosion producing a radioactive leak) at Chernobyl. Fortunately, radioactive decay (half-life) of the escaped products was very short in that case, but a year later shoppers were still going to the market with radiation detectors in hand.

In existing arsenals the capabilities for deliberate destruction already far, even massively, exceed the permissible limits. When the planners talk about hundreds or even tens of nuclear explosions, they are indeed playing computer games. To be slightly uncouth, it is an exercise in mental masturbation to suggest to any population that you are protecting them but will allow a few tens of nuclear explosions to occur in the process. Thus any defensive system that from the outset will permit impacts in excess of some tiny limit of nuclear explosions (of the order 1.0) actually does no good at all except from the traditional military viewpoint.

The military desires protection of the retaliatory capability believing that increases deterrence and thereby increases security of the population. But from the population and national assets point of view, either the protection is 100% effective from a nuclear "rain" or it is nothing. In fact, to use numbers such as 99% effective is worse than nothing. You might be lulled into thinking that 99% is rather effective or that we just might survive 50 or 100 nuclear impacts in the U. S. (or the Soviet Union), but if the threat is 10,000 nuclear impacts, not even the most foolhardy will take the risk of setting it off. With the continuous increases in the number of weapons with nuclear capability, the problem is already far beyond what is even possible to think about in terms of limiting population damage, should a war begin.

There are not enough 999s attainable in the effectiveness percentages to defeat by technological means the escalation of numbers of weapons. The foregoing is true whether applied to nuclear weapons or the new technology systems which might be added to enhance or counter them. It is precisely in this argument that the different risk factors between population and their champion asserts itself. Each evaluates the problem from a totally different point of view.

3.1.2—THE MILITARY VIEWPOINT

In the councils of government we must guard against the acquisition of unwarranted influence, whether sought or unsought, by the military industrial-complex. The potential for the disastrous rise of misplaced power exists and will persist.

— Dwight Eisenhower, Farewell Address, 1961

From the military point of view, the argument about the numbers game is a bit different. There is no policy regarding the numbers game except militaries always would like superiority. Superiority as an official policy has been renounced but it is relentlessly pursued unofficially. Increasing the required numbers just adds power and prestige to the military and increases its budgets. There is no policy incentive to reduce the numbers. The announced policy and strategy is deterrence. The argument is: "An adequate defense at all costs." But "adequate" should be read as "superiority".

When it was originally hoped that population defense was possible, that hope formed the rationale for SDI. Now that population defense it seen to be infeasible the argument has reverted to enhancing deterrence. Deterrence is designed to reduce the likelihood of a deliberate attack. The strategy for this contains two elements: 1) maximum protection of retaliation capability and 2) be able to destroy with SDI a sufficient number of the attacking missiles so that the opponent is less certain that his attack can be successful. Retaliatory capability and uncertainty, it is argued, is each an important element of deterrence. The more of each the better. The threat of retaliation is ultimately the deterrent; and uncertainty is said to enhance that deterrence. Before the current situation, that traditional argument would be valid. It ignores, however, the eventual consequences that the escalating technology produces.

By its definition deterrence implies that the opponent would launch a deliberate attack under certain circumstances. There can only be two such circumstances: he has aggressive intent or in self-defense. Deliberate, rational attack with aggressive intent has near zero probability of occurring. The destructiveness of nuclear firepower has precluded that eventuality. That would only happen under the old rules of the game before ICBMs and massive overkill capability. No sane political leaders on either side are going to use strategic nuclear weapons for a deliberate aggressive attack. They will not do so if there is a likelihood that even a small number can be returned to fall on their population and national assets. Thus enhancing deterrence of a deliberate attack by trying to create uncertainty for the opponent through SDI is a meaningless exercise. It is an exercise in erecting straw men in order to knock them down. What purpose is served by trying to create greater assurance that the opponent will not do what he already has no intention of doing? Aggressive intent must be expressed in other ways than nuclear attack in this era. And if the opponent is just defending himself, he does not need to be deterred. But nevertheless, that is the argument. The existing rationale exhibits a paranoid preoccupation with deliberate attack while ignoring other more vital elements of the issues.

These arguments used by proponents of SDI assure that escalation in numbers and technological innovation of measure and countermeasure relentlessly increase. It is a game that has no technological solution. The only solution to the threat of increasing amounts and variety of weaponry is a decrease in the amounts of weaponry. The particular technology itself is another problem quite apart from the numbers of individual pieces of that technology that one employs.

Let's now examine the technologies that are likely candidates for the SDI system. It is important to understand what these weapon technologies are in order to decide whether their use enhances or detracts from security and stability and by how much. Are they valid areas to pursue in that their use tends to stabilize things and reduce world tensions or are they inherently destabilizing in armed confrontations? The numbers game is a difficult game by itself, but it is not the only aspect of the issues relevant to SDI. There are technologies such as observation and communications technologies that do promote stabilization. There conceivably could be others.

The new areas of knowledge being considered for SDI fall into three distinct categories: radiation, particles, and information management. There are also some older technologies which, with new basing modes and engineering improvements, need to be examined as well. These are called kinetic energy weapons, i.e., projectiles of various types. In the early 1980s they were not considered suitable for space use; now they are being resurrected for certain parts of the system.

3.2—CHEMICAL LASERS

There are more things in heaven and Earth, Horatio, than are dreamt of in your philosophy.

— Shakespeare, *Hamlet*, Act I, Scene 5

A development since World War II has been the laser. Lasers are just machines that emit light beams in a particular way. Whereas ordinary light from the sun or light from artificial sources is made up of many frequencies and is unpolarized, laser light is not. Laser light may be thought of as light that has "combed its hair." Instead of having the wave forms oriented in random directions like an uncombed head of hair, the waves of laser light are all coherent, i.e., organized, so that all the waves are in step like troops marching. All waves are of the same or of a very limited number of frequencies. Certain materials can be electrically stimulated in ways that cause them to emit radiation in this fashion. Other materials can be burned to produce coherent radiation. A property of this radiation is that it is easily focused into precisely controllable beams. Another is that this permits a maximum amount of energy to be transported and focused in a very small area, much as a magnifying glass can focus sunlight to burn a hole in a piece of paper. Already lasers have found their way into medicine and into industrial and scientific use where they are enormously beneficial in numerous applications. But these beneficial lasers are tiny and very low powered compared with what is being considered for the future.

The potential weapons application of lasers is as a device to burn through or destroy material structure (including people's bodies). It does so by delivering sufficient energy into the atomic structure of the target to vaporize the material through localized heating and/or high energy impact. In applications which can destroy with surgical precision small areas of cancerous tissue or burn small precise holes in heavy metal materials, the technology is very well in hand and in useful daily application.

What is not proved is how to build giant laser systems that can reliably send beams across thousands of miles with enough focused energy to destroy satellites in orbit or missiles rising in ballistic trajectories under battle conditions. This latter is a problem many orders of magnitude more difficult than just constructing a simple laser. It is not just an engineering problem since many fundamental questions remain as to how materials will behave under such operating conditions. Depending upon the type of laser, this latter application requires peak power of the order of gigawatts.

There are a number of ways in which radiation can be stimulated and focused into a laser beam. The techniques vary depending upon the frequency of the laser radiation desired and energy required in the beam. All techniques depend upon exciting the electrons of some substance to change their position in the molecular structure of the substance and thereby release radiation of the desired frequency. Various chemicals are selected according to the ease with which the desired frequencies may be obtained to the exclusion of other frequencies.

Developments in quantum theory of subatomic particles during the past 40 years have given rather precise knowledge as to how matter behaves and how it can be manipulated at these levels of its existence. At the level of atoms and molecules there are no longer many surprises. Most of the behavior is known at least in principle. The lasers under discussion utilize atomic knowledge and therefore are problems in applied science rather than basic science. This distinction is important because it tells us that the devices under consideration are feasible and that their gross characteristics are predictable from known physical laws. Only time and resources are required to eventually construct such weapons.

One of the candidate weapons for the Strategic Defense Initiative has been a chemical laser. The applicable types would burn either hydrogen fluoride (HF) or deuterium fluoride (DF). Both are candidates for space basing. These lasers are known as continuous wave lasers in that they emit energy in continuous beams rather than in short intermittent pulses. The energy is obtained by spraying the appropriate chemicals together in a mixing area where the light energy from the chemical reaction is extracted and focused by mirrors.

HF laser light can only penetrate a certain distance into the atmosphere before being absorbed by water vapor and other molecules in the atmosphere. It is potentially most useful, therefore, against satellite targets or aircraft and missile targets above the troposphere, of the order of 15 kilometers above the surface. It is increasingly absorbed as it travels from space deeper into the atmosphere and only small amounts of energy would reach the surface.

The atmosphere is transparent to the frequencies of the DF laser, thus it could be used against targets on or above the Earth's surface. The HF and DF lasers would have many of the same physical and operating characteristics. The primary operating distinction between the two is the greater depth down into the atmosphere that the laser energy can be deposited in a coherent beam. Atmospheric losses affect all laser light through a number of interference mechanisms, but those lasers emitting light to which the atmosphere is transparent can deposit significant amounts of energy down to the Earth's surface. Being continuous wave machines, once activated they emit energy much like a searchlight and deposit energy over a period of from one to several seconds on the target to effect a kill. To kill hardened ballistic missiles with this type of device requires laser power of the order of megawatts at the source.

The size of such a weapon and its control machinery is estimated to require a space roughly equivalent to the size of 747 aircraft of the order 10,000 meters cubed. Its weight would be of the order of a hundred tons plus fuel. Any systems to defend the installation would increase the size and weight. The most conservative early estimate (by advocates) of the number of this type of laser required in space to counter the existing Soviet ICBM installations has been 90. The number increases to several hundred as different technical and strategic assumptions are made about the battle problem.

The foregoing is just the minimum initially estimated. The chemical laser, although originally given considerable emphasis as a candidate weapon, has been downgraded for use with SDI in the near term because of the complexities associated with its development. However, in the longer-range escalation of measure and countermeasure it must be considered very potential as a weapon because of the ability to be space-based. Not all candidate weapons for SDI have a potential for space basing and at some point, should SDI proceed, the chemical laser will return as a priority item. There are, however, a number of fundamental technical problems associated with this type of weapon which need research and thus are worth mentioning in this section.

The first major problem is the mirror. Lasers of lower frequencies, particularly near the visible spectrum, use a mirror as a device to focus and direct the beam. Medical and industrial lasers use lens and mirrors a few inches in diameter. Those under consideration for missile and satellite killers begin with mirrors about ten meters (thirty feet) in diameter and increase to many times that size. The mirror must be exact in its shape to the order of a wavelength of the laser beam frequency. It must be free from defect and contamination in order to reflect, rather than to absorb the laser energy. A cup of oil thrown on the surface would, for example, destroy the reflective characteristic, cause the surface to absorb energy instead of reflecting it, and thus destroy the mirror instead of a missile.

Such mirrors must be almost 100% accurate and free of distortion. The size and exactness requirements create vulnerability of mirrors to micro-meteor hits which can damage the surface while in space over long periods of time. The need for near perfection in the mirror surface also offers a very simple countermeasure against it. Almost any device can be used to destroy the mirror's reflecting surface. These are problems, but not insurmountable ones, for a peaceful space environment. The countermeasure issue, protection of the mirror from deliberate destruction attempts in war, seems nearly impossible, however.

The next major consideration is vibration control and pointing accuracy. The vibrations of the chemical laser affect the shape and pointing accuracy of the mirror system. Lasers of the size considered for weapons are noisy and vibrate about the same as a very big jet engine. (They burn large amounts of chemical substance to produce the light and it will be pumped through at supersonic speeds. Fluids moving at supersonic speeds are noisy and difficult to manage.) Such vibrations must be completely isolated from the mirror if the mirror is to function properly. In space, isolating vibration is almost an impossibility because there is nothing to serve as an anchor point. The problem is contained by using massive structures which absorb the vibration. (The problem is like wagging the dog but not the tail.) On Earth, large blocks of cement isolated from the surroundings create vibrationless environments.

This is obviously not feasible in space. Massive structures isolated from each other by some damping medium must be used, however. And the need for massive structures itself poses a large difficulty for space systems. Size, mass, rigidity, and vibration damping are trade-offs that are an inherent weakness because 1) launch weight is a problem, because 2) every action causes an equal and opposite reaction in the frictionless environment of space, because 3) fuel must be consumed to maintain the system in its stable position when the laser fires or vibrates.

Although mirrors with the necessary size, smoothness, and accuracy characteristics are not considered infeasible, managing them in space under battle conditions is an enormous problem. They must be controlled to point accurately on one target from tenths of a second up to no more than a couple of seconds, then shift to the next and stabilize any oscillations between each laser shot. Moving such large devices quickly without surface distortion and vibration to point accurately on targets even a hundred miles away is not insignificant and particularly not so from a space platform. And the SDI proposals bring the need for accurately pointing over several thousand miles. The nearest Earthbound equivalents are the large mirrors of the astronomers' telescopes. Think of the size, weight, and cumbersomeness of these devices, then try to picture even larger mirrors being accurately managed in space in an attempt to locate and shoot a different missile every few seconds. It is a difficult but not unattainable feat in principle under stable conditions. In wartime battle conditions, however, the problem assumes a different level of complexity.

A likely solution to some of the mirror problems in space will be sought through use of smaller mirrors controlled as a "phased array." Many mirrors properly arranged and controlled by computers and mechanical actuators can be made to operate like a mirror many times the size of the small ones. For the very large mirrors of a space-based system, the phased array solution would seem the most likely.

A few meteor strikes, a tiny amount of oil or chemicals, or an aggressor's activity can reduce the mirror effectiveness quickly to zero. Most of these problems would be insignificant in an Earthbound maintenance environment. They become huge obstacles in space. Unless such seemingly small issues as these are solved, however, the laser and its mirror in space become just another piece of expensive space junk in spite of its exotic technology. Protecting the system and maintaining operation while under attack appear as almost insurmountable problems because of the hostility of the environment and delicacy of the machinery.

Another major operational problem of chemical lasers is fuel consumption. Like an engine, the fuel consumption is dependent upon the necessary power output. The number most frequently used for the proposed system is about 100 kilograms (200 pounds) of chemicals to produce the energy to penetrate a thin-skinned target. This amount would be burned in approximately one second, thus a fuel flow of 200 pounds per second. This is about an order of magnitude more than required by a large jet engine at full power. To kill hardened targets would require perhaps ten times as much energy as a thin-walled missile, requiring ten seconds instead of one, for example. The fuel required in space then depends upon the number of shots that each laser is expected to make, and the total number of stations required in the system. To shoot 10,000 times for 10 seconds each time would require 10,000 tons of fuel in space. Early estimates of the fuel required for the total system begin at the number 10,000 tons of chemical fuel and increase to about 60,000 tons. The efficiency of the laser dramatically affects the amount of fuel required. To give an example of the severity of this problem, a laser system designed within the foregoing parameters would require the order of 1000 of the existing shuttle payloads to put the fuel for the system in place.

The final major problem is the laser itself. It undoubtedly can be constructed in due course. The report to the American Physical Society in April, 1987, stated that power output would need to be increased by two orders of magnitude (100 times) over existing devices to have effective weapons for missile kills. Without knowing the status of classified research it is still very easy to project that there are numerous undiscovered problems with regard to controlling devices of this size and energy output. The complications are approximately equivalent to building a jet engine of the same size. It has required 40 years of constant improvement in jet engine technology to achieve the levels of energy output enjoyed today. There is little reason to believe that reliable orbiting lasers of this size can be perfected in significantly less than one-half that time. Particularly, it is unlikely that they can be perfected to the point of unattended operation in the space environment in the very near term.

A problem of space basing of any device which operates in the power range of many megawatts up to gigawatts is cooling. This is true of lasers, power supplies, or any other power consuming device. All will operate at efficiencies of only a few to a few tens of percent. The excess heat must be managed to keep components from being destroyed from overheating during their normal operation. In the space environment this is not a simple problem. Unlike the Earth environment, of course, no plentiful substances such as air or water are available to provide natural cooling. Thus self-contained artificial cooling systems relying upon high temperature heat radiation to remove excess heat must be used. The laws of thermodynamics are relentless in this regard and again pose significant problems for space systems designers. Recent breakthroughs in super-conductor technology will help this problem for electrical devices but as yet can do nothing about thermal processes.

With chemical lasers, the gaseous residue from the process can be exhausted into space and thereby carry away the heat. But in so doing, two new problems arise: the first is a gas cloud around the battle station through which the laser must "look" in order to shoot. The second is the destabilizing effect of the gas exhausting from the laser which requires fuel to re-stabilize the battle station. These again are manageable in prin-ciple but add to the complexity and weight of the system.

Although lasers are vitally important for the success of any futuristic ballistic missile defense system, the aforementioned problems of weight, mirror size, vibration, fuel consumption, cooling, all pose significant handi-caps for early space basing of equipment in addition to the develop-ment problems of the lasers themselves. It clearly will require quite a number of years of development and test. Such problems are not insolv-able, however, and will in due course of time succumb to engineering genius. But for these reasons, other types of devices and configurations of equipments are under consideration for more immediate use. One of the alternatives is to base lasers on the surface and to put only the mirrors in space to reflect the laser beam toward the target. Another is to use the laser on the surface as a terminal phase device against missiles or as an anti-satellite device without using space mirrors at all.

For SDI to meet any of its objectives, however, some devices must be placed in space. The "numbers game" is such that attacking oncom-ing missiles must begin at the instant they are launched and continue throughout their flight to the target if any statistical measure of success is to be achieved. The flight times of nuclear missiles, being less than 30 minutes, are too short otherwise. Because launch vehicles can each carry numerous warheads and each warhead can be protected by numerous decoys, the acquisition, tracking, and targeting problems individually become an order of magnitude more complicated at every stage following launch, through midcourse and into the terminal phase.

The geometry of the problem dictates therefore that each kill in the boost phase eases the problem of later phases by at least a factor of 100 and perhaps more. Disposing of one launch vehicle ensures that up to 100 warheads or decoys would not need to be tracked and destroyed in a later phase. By any measure of effectiveness, if the problem can be solved at all, one must begin the solution by trying to dispose of missiles as they rise from the launch pad. This can only be accomplished with a space-based system.

Assuredly then, even though the most effective and "clean" missile killer, a laser, may take years to develop, if defense against nuclear missiles is to proceed, the end point will someday include lasers in space. Until that time, however, other systems and partial systems are being developed.

3.3—SURFACE-BASED LASERS

My candle burns at both ends;
it will not last the night;
But, ah, my foes, and oh, my friends—
It gives a lovely light.

— Edna St. Vincent Millay, *A Few Figs from Thistles,* 1920

An immediate advantage obtained from basing lasers of the necessary size on the surface derives from the fact that the weight, power, vibration, and maintenance requirements are infinitely easier to satisfy. If the total problem of space basing is too difficult to solve immediately, it seems obvious that the lightest, least complicated part of the system should be considered for launch into orbit first and the most complicated part be based on the surface where it is easily accessible for fuel and maintenance needs. Surface basing of lasers, at least initially, also permits development to proceed more easily toward perfecting weapons which can eventually be orbited. This approach suggests ground basing the laser and only orbiting the mirrors. Although the mirrors are a complex problem, they are less of one than the total system taken altogether. This approach, however, brings new and additional problems which are also severe. The mirrors sizes in space now must be even larger and there must be more of them. The laser beam must be bounced off a huge mirror far out in geosynchronous orbit which in turn bounces the beam off a low orbit mirror toward the rising missile target.

This has been described as a 50,000-mile, two-cushion billiard shot.[15] The lower orbit mirror must now be about 50 feet in diameter for HF and DF lasers, and the controlling mirror in deep space must be almost an order of magnitude larger to do its job. There must be one mirror in deep space for each laser and a number of "kill mirrors" sufficient to focus on all targets. These huge mirrors would likely need to be phased arrays in order to solve all the problems that such an approach brings.

For surface basing, however, a different type of laser is considered more suitable. It is called an excimer laser. Instead of burning chemicals it uses excited electrons of rare gases such as krypton, argon, and xenon compounded with fluorine or chlorine. These would operate at natural wavelengths an order of magnitude shorter than chemical lasers but can be designed to achieve a number of different wavelengths corresponding to "windows" in the atmosphere by shifting the natural wavelengths. This can be done by passing the beam through a gas which emits the correct frequency. Because they are ground based, the additional weight and complexities which preclude excimer lasers from orbit are of little concern. The shorter wavelength of the excimer laser also increases its kill range by a factor of ten over chemical lasers using the same sized mirror or alternatively permits smaller mirrors for the same kill range. This latter attribute eases the mirror-in-space problem considerably, bringing them down to more reasonable sizes.

The excimer laser is a "pulsed" laser rather than a continuous wave device. It delivers its destructive energy in very short pulses (of much less than a second duration) but of very high power in each pulse. Peak powers in the gigawatt range are required to drive this device. In addition to thermal vaporization of a target, the energy pulse is sufficient that the impact momentum of the photons of light play a role in destruction of the target as well. Pulsed lasers can kill a hardened missile target in less than a tenth of a second whereas a continuous wave laser might require several seconds illumination but at lower peak power.

There is, however, a significant energy loss problem with lasers even though "tuned" to atmospheric windows. A certain amount of energy is absorbed and scattered simply as a result of encountering the moving molecules of air and water in the atmosphere. Tuning can never be so perfect that all losses are avoided. As a result, there is a certain amount of heating of air molecules that will take place causing convection currents in the atmosphere in addition to normal wind currents. Thus diffusion and scintillation (twinkling) will tend to take place, the same as with visible star light. This would tend to make the laser light become incoherent. The former is mostly compensated by using more energy in the laser and accepting the losses. There are proposals for eliminating the scintillation by having the laser mirror actively change its shape to compensate the "twinkling." Unlike space-based lasers, a ground-based laser can afford the additional weight, complexity, and inefficiency to "blast its way" through the atmosphere in order to deposit the necessary energy on a target or a reflecting mirror.

Initially it was believed that these atmospheric losses produced such a severe barrier for lasers in space that little energy would penetrate through the atmosphere to the surface. Such a device could not effectively reach the ground to pose an offensive threat to surface targets. On this basis a chemical laser in space could be labeled a "defensive weapon." Only four years have elapsed since the initial SDI proposal and already laser light is being "shifted" to match the windows in the atmosphere and compensate for turbulence so that laser beams can pass through with little absorption or disruption. The issue in this case is no longer technical but strategic. If the laser light can go outward from the surface through the atmosphere and be a satellite or missile killing weapon, then it can also come back from space to the surface and hit a target. This can then be an offensive first strike laser weapon as well as a defensive one. The last few miles of atmosphere not initially considered penetrable can be penetrated. Wave lengths of laser radiation tuned to the windows in the atmosphere can pass through in either direction not withstanding that losses and distortions occur. The problem is one of emitted power of the laser, focus of the mirrors, and frequency of the light. Hardened sea-level military targets may not be vulnerable to such devices, but the soft assets of the population would be.

Defensive technology is thus offensive technology which will hit the surface sufficiently to be destructive. In fact they are weapons of mass destruction having the capability to create large firestorms on the surface against soft and flammable targets. This then is the crux of the laser problem as presented in this book. The laser capabilities bear out the earlier argument that no weapon can be considered only a defensive weapon. They are all just weapons. How they are used is determined by the intent of the user. With both sides having such devices, defender and aggressor still appear exactly the same. The problem for each is reduced to measure and countermeasure rather than attack and defense; but with zero warning time. With the destructive energy traveling at the speed of light, ultimate surprise is thus possible!

To place numbers on the foregoing population threat, consider the following. The kill energy needed for missiles and nuclear warheads is computed in the range of 20 to 60 kilojoules/centimeter squared.[16] A kilojoule is the energy of a kilowatt of power delivered in one second. Thus continuous wave lasers would need to deposit energy at the rate of 20 to 60 kilowatts/centimeter squared. The sun's energy when directly overhead is .13 watts/centimeter squared at the Earth's surface. Even if the atmospheric losses of the laser were so great as to reduce the energy flux by a factor of a thousand or more, its energy would still be 10 to 1000 times brighter than the sun, sufficient to fire flammable material. This would be the side effect on Earth of missile killing lasers in space. But a "defender" determined to "enhance deterrence" could increase the laser power to pose direct and even greater threat to population assets.

Another type of laser which has been suggested and which avoids some of the issues of the chemical and excimer lasers is the X-ray laser proposed by the Lawrence Livermore Laboratory. This system would not use mirrors for focusing because the wavelengths are too short. Mirrors are totally infeasible for this type device. Focus must be achieved by physically pointing the device which creates the lasing medium, in this case long thin rods of metal which are vaporized serve the purpose of focusing and pointing. The X-ray laser also avoids the vibration problem, is relatively light weight, and avoids much of the fuel and machinery problems because the proposed fuel source is an exploding nuclear bomb. Another advantage, at least theoretically, is that it should be able to have multiple "lens" to point at several targets simultaneously.

The energy from any one nuclear explosion is sufficient to kill several missiles if the energy can be correctly focused on several of the targets simultaneously. Also, and very importantly, X-rays do not penetrate deeply into the atmosphere, thus there is no opportunity to hit aircraft or ground targets. All targets must be in space or near space to be destroyed by an X-ray laser. Such a devise cannot be used except near the end of the launch phase after the missile gets above the atmosphere. It is a good midcourse phase weapon and a particularly good anti-satellite device, however. Those are the advantages; now for the disadvantages.

With this system, either the nuclear bomb fuel supply which pumps the lasing medium must permanently orbit in space or it must be launched into space at the time of an attack in order to counter the aggressor missiles. The latter technique is called a "pop-up" delivery method.

Orbiting nuclear weapons in space for "defense" not only violate existing space treaties but is a heinous concept, which if employed, could only lead to an escalation in which orbiting satellites routinely carry nuclear weapons for "offense." This solution of having nuclear weapons orbit in space would only transplant the existing "deterrence by threat of retaliation" stand-off with an even more risky and tension-filled orbital situation. In such a case, warning times decrease to the time necessary for de-orbit, and that is of the order of a minute or two. We shall not discuss this option further because it is not likely to be seriously considered for SDI and because it is so ridiculously destabilizing that no reasonable individuals would propose it for a defensive system.

The thought of orbiting nuclear explosives is so threatening that it needs no explanation to the public. The answer will correctly be "no" regardless of the circumstances. The MAD environment would become even more untenable than the one now in existence if both opponents resorted to this approach. However, if SDI were even to use the pop-up mode of delivery, the next step in the endless escalation of weapons technology would surely see attempts to justify orbital nuclear weapons in order to maintain superiority. Once the breach of space with weaponry begins, there will be little hope of stopping another massive round of escalation in all its varied forms.

Thus, for the X-ray laser and its nuclear bomb fuel supply, launching at the time of an attack in order to counter missiles or satellites which have already been launched offers the only realistic option for this candidate weapon. The launches would likely be from submarines positioned as near as possible to aggressor territory in order for the laser to arrive on station in time to be effective. The existence of the nuclear bomb fuel supply, even though not planned for launch until an attack has already begun, is destabilizing from the point of view of managing global conflict. Its very existence encroaches on the treaty banning space nuclear explosions and invites countermeasures escalation to include other nuclear devices in space.

Because the atmosphere is opaque to X-rays and the target must be near the top of the atmosphere before an X-ray laser can become effective as a counter-weapon, the problem of acquiring ballistic missile targets for the X-ray laser is much more difficult. The fiery tail of the aggressor launch vehicle is gone, and the incoming warhead is now a "cold" target. Cold targets require radars for tracking instead of the passive infrared sensors normally used to detect the exhaust plume of a launching missile. Further, the fact that both the pop-up laser and the missile target are on sub-orbital powered trajectories rather than orbital trajectories enormously complicates the acquisition and targeting problem. Sub-orbital ballistic trajectories, particularly while still experiencing atmospheric "drag," are much more variable thus less predictable than orbital trajectories. Rising from under the water to 100 kilometers above the surface, acquiring, aiming and firing within about 90 seconds while both laser and target are on ballistic trajectories is quite a formidable acquisition and computational task.

The very limitations which the X-ray laser must overcome to kill nuclear tipped missiles make it a very good weapon for killing satellites in orbit, however. If the satellite position is known from other means, which it always is, the laser can pop up and reach across long distances of space to make its kill based upon information passed to its computer from a tracking net. The speed of the action slows down for the orbital problem but is very time critical for the ballistic missile problem.

A major problem with any nuclear weapon exploded in space, such as the X-ray laser, is the effect on the atmosphere. Tests have indicated that high altitude explosions of nuclear weapons shower the atmosphere with particles and radiant energy which disrupt electrical systems, radars and communications systems over a very wide area of perhaps several thousands of miles.

The effects have not been totally investigated, although the principles are completely understood. A number of X-ray lasers fired at missile targets would likely disrupt all forms of electrical systems, causing large problems for tracking radars and Earthbound or space battle management systems. Some authorities believe that exploding nuclear weapons in space over or near a country will have such disruptive consequences that virtually all communications and electromagnetic devices will be disabled for a significant period, if not totally destroyed outright. The micro-circuitry of computers and electronic equipment is particularly susceptible to overload by such electromagnetic pulses.

A fourth type of laser is also under basic investigation. It is called a free-electron laser.[2,15,16] Its principle is that of passing high-speed electrons through a non-linear magnetic field in such a way that the electron is induced to give off radiation of the desired frequency. In principle any desired wavelength of laser radiation may be obtained in this fashion. Thus, the various "windows" in the atmosphere transparent to radiation can be penetrated by selecting the proper wavelength for the free-electron laser. This system is less well developed theoretically than other type lasers but holds great promise because of this potential to select from a continuous range of wave lengths. The system requires an electron accelerator of greater power output than has been built thus far. It is not a good candidate for space basing because of size, weight, and power requirements and very low power efficiencies. In this regard the free-electron laser resembles the excimer laser. It might not be capable of being based in space but would be capable of killing satellites in space, or as a possible terminal phase device. Through use of reflecting mirrors, it can be capable of striking targets on or above the Earth's surface anywhere in the world.

3.4—PARTICLE BEAMS

Life is a wave, which in no two consecutive moments of its existence is composed of the same particles.

— John Tyndall, *Fragments of Science*

Another theoretical system that has been proposed uses particle beams such as electrons, protons, neutrons, and hydrogen atoms as bullets instead of the photons of radiant energy. These are highly speculative systems, even more so than the others. The shortcomings immediately known for these systems are many. First of all, charged particles repel like charges, thus they disperse in space instead of staying in a well-focused beam. They behave like a free flowing electrical current and thus strongly and unpredictably interact with the Earth's magnetic field. Because of this they cannot be pointed accurately over long distances without some auxiliary device such as a laser creating a channel for the beam. Neutrally charged particles like neutrons and hydrogen atoms would not have some of the foregoing problems but are significantly harder to produce than other type particles. The complexities of producing them will be studied, however, because neutrally charged particle beams do have a short range kill capability that is different from lasers.[2,16]

Whereas lasers deposit the photons of energy in the beam on the surface of a target and destroy by thermally evaporating successively deeper layers of the skin, heavier particles like protons immediately penetrate more deeply into the interior of the object and release their energy within the molecular structure. For dense and heavy targets such as the ignitors for nuclear warheads or for hardened projectiles, particles are a preferred method of destruction. Also for interrupting and destroying electronic circuitry, particle beams are a superior weapon. If used against command-and-control systems or battle management satellites, the penetrating capability of the particle beam is devastating to computer and other electronic devices and at power levels much less than those required for missiles. It would be unnecessary to vaporize all the material surrounding the electronic devices before destroying the device itself. Lasers, on the other hand, must burn their way through a hardened target before being able to destroy sensitive equipment inside. For this reason and at ranges less than five or six hundred miles, particle beams can potentially be a very powerful weapon.[2,16]

A significant fact about particle guns, however, is a side effect of the property that makes heavy particles effective against electronic components. It is the same problem as with the nuclear-powered X-ray laser, where the shower of particles into the atmosphere from the nuclear explosion sets off secondary radiation by interacting with air molecules. The electromagnetic effects of particle interactions can cause serious disruptions of electrical power systems and communications systems in areas hundreds of miles distant.

The particle gun would require power on the order of several gigawatts for missile kills but much less for electronic disruption. This weapon, like the X-ray laser, is a candidate for countering warheads in the midcourse phase at short ranges, against satellites, particularly battle management satellites and as a counter to kinetic energy weapons aimed at satellite stations. These modes clearly require space basing. It is not likely that this type of weapon has any future in attempting to stop nuclear tipped missiles rising from a launch pad because of the Earth's magnetic and atmospheric effects.

All the foregoing weapons based upon different types of lasers, beams and particles can be developed in principle and placed into use in the future, with due regard for their individual limitations. Each has certain strengths and weaknesses, a characteristic of any technology. Clearly, from the current state of the art, development will require significant time, effort, and resources, and improvements by several orders of magnitude over existing devices. Additional years will be required before operational use of such systems would become feasible. The managers and proponents of SDI now recognize how formidable is this task. The arguments presented in this book, however: that the end point, even if achievable, is totally undesirable and self-defeating from the point of view of stabilization of global conflict and world peace, have been totally swept aside. There is a concerted effort to commit some systems to operation before the end of the Reagan administration and thereby make it more difficult to slow down or stop the total process of getting weapons into space.

Toward this latter end, more mature and simple technology is being pursued because it will require less development time. The immediate priority of pursuing kinetic energy weapons is the result. Even though they are only a step along the way toward the instant kill capability of lasers, kinetic energy weapons are almost as destabilizing as lasers and create problems of their own with regard to decision time. Actually it is the circumstances of deploying a space weapon for boost phase kills that reduces decision time and not the particular weapon that is employed.

Thus most every argument that is made herein about exotic weapons designed to kill rising missiles is equally valid for kinetic energy weapons. They are just a slower and less reliable means of stopping missiles, but they equally produce a condition of instability.

3.5—KINETIC ENERGY WEAPONS

Though boys throw stones at frogs in sport, the frogs do not die in sport, but in earnest.

> — Bion, as quoted by Plutarch, *Moralia*, xii. 66

Because of the high velocities of objects in space, of the order of 8 kilometers per second (25,000 feet per second) for low Earth orbits, the kinetic energy of impact between objects can be very destructive. The "well" of the Earth's gravitational field requires such velocities to overcome the centripetal attraction and remain in orbit. If the geometry of the impacting objects' trajectories are opposing each other the velocities are additive. The destructive energy increases as the square of the closing velocities. Thus fragments from an explosion in front of a target would take advantage of the kinetic energy in the respective motions of the objects. The energy from such impacts at orbital velocities can be even higher than that deposited by a killing laser beam over short time periods. All the energy of impact is delivered at once rather than distributed over a few seconds. The damage is thus explosive. A one-gram mass impacting a satellite target in this fashion could penetrate about 15 centimeters (6 inches) of steel. This is about the amount of armor plating that a battleship carries. A pattern of such simple projectiles would act like bird shot from a shot gun but with sufficient energy to blow a large hole in armor plating. Because of the weight problem, however, satellites can hardly be armor plated for their protection to such an extent.

Kinetic energy concepts are not really new ideas. The stones thrown by "the boys at frogs" were kinetic energy devices. They now just receive a new name more in keeping with the space age. It is precisely the high speed, thus high kinetic energy, combined with predictable orbits that make space objects so vulnerable to destruction. The energy imparted just by an object much smaller than a golf ball will explode through a satellite more destructively than an armor-piercing shell through a tank. It is for this reason of power and simplicity that kinetic energy weapons command a high priority in SDI planning, about 25% of the current budget. These weapons would be the least expensive and require less immediate breakthroughs in technology. The problems with these devices, as compared to lasers, are associated with their low speed relative to the speed of light. Whereas a laser can reach across thousands of miles of space to kill a target almost instantly, kinetic energy weapons have a velocity of the same order as their targets. They require a significant time to reach the target so that acquisition, tracking, and aiming, although more conventional, are also more difficult computational problems.

For the terminal phase of interception, i.e., as the incoming missiles descend toward their targets, surface-based kinetic energy weapons are the most immediately promising. Work has been progressing for years on these weapons just as a normal extension of existing ballistics and rocket technology. Surface-based terminal defense with missiles is by far the most mature of the technologies available to SDI. But even though tests have already demonstrated the capability to intercept an incoming test target it is quite clear that terminal defense alone could never satisfy the objectives of SDI. This would be true regardless of the technology involved because saturation of terminal defenses would quickly occur with even the most rigorous and rapid firing installations. The terminal phase of an incoming ballistic warhead as it descends into the atmosphere is of the order of one minute. The existing 10,000 warheads plus at least ten times that many decoys would pose an insurmountable problem if left to terminal defense alone. The "numbers game" would surely win if countered only in the terminal phase. Thus the midcourse phase above the atmosphere and the boost phase must be used to attain an effective number of missile kills if there is to be a defense at all. The terminal phase should be relied upon only to "clean up" those missiles that escape being destroyed in the boost and midcourse phases.

Terminal phase installations for defense of particular targets, for example, the Capitol, are called "point defense." These are ground-based installations and could be consistent with the ABM Treaty. The current Treaty limitation is one point defense installation on each side. The Soviet has chosen to deploy its one allowed site around Moscow. The U.S. chose to dismantle their one allowed site. It was not considered worth maintaining. To surround additional sites with such defenses requires a modification of the ABM Treaty. One protected site, however, or even a dozen protected sites does little for defense in general and absolutely nothing for the population. If there is to be a defense at all, it needs be perfect and no amount of terminal defense will ever satisfy this need.

Turning attention to midcourse and boost phase defense with projectiles, the problem is to get speeds and guidance accuracies such that projectiles can reach their targets from space bases. The need is to have intercept ranges compatible with the space environment and that is of the order of at least 1000 kilometers or about 600 miles. Normal intercept rocket technology can serve in some instances for basing in space, but higher speeds are also desired in order to reach the target more quickly. Interceptor rockets can easily achieve velocities of the order of 10 kilometers per second, but velocities about twice that are preferred. A propulsion system that might accomplish this is called a rail gun. It likewise is not a new concept. It is an electromagnet propulsion system that can achieve very high accelerations and thus very high velocities. A simple example of this type of device is a common household solenoid which moves a metal plunger which then rings the doorbell when the current is applied. With projectiles accelerated along the length of a 200-to-500-meter electromagnet "rail" solenoid, very high velocities are obtained.[16] When projected into the path of an oncoming satellite or warhead, the destructive energy would be enormous. This type device could project a "bird shot" pattern of inert projectiles in front of incoming warheads. It could also intercept rising missiles at the top of the atmosphere. Deep in the atmosphere where the missile is rising more slowly and the atmospheric drag would also slow down the projectiles, a "bird shot" technique would be less effective.

Guided projectiles and explosive projectiles propelled by either a rail gun or by normal rocket technology can find use in each phase of the defense system. The most likely guidance technique for projectile systems centers around infrared sensing devices which detects the rising missile's exhaust plume. These would only be applicable, of course, during the boost phase. But if the defending projectile penetrates too deeply or too quickly into the atmosphere, its own heat of reentry "blinds" the guidance sensor. Thus there is a delicate task of computing when and where the intercept and kill should be made. If the rising missile is allowed into space and its booster shuts down, then the problem changes. Different sensors are then required for detection and tracking plus the fact that a number of warheads and decoys have likely been deployed from the launch vehicle. The "window" available to the kinetic kill device to intercept a rising missile in its boost phase is of the order of 20 to 40 seconds. If that opportunity is missed, the intercept problem becomes enormously more complex by an order of magnitude or greater.

During the boost phase the target is the rising missile's fuel supply, protected only by a large relatively thin-walled fuel tank. By penetrating the tank and exploding the fuel, the kill is simplified. After the launch vehicle burns out and separates, where only the smaller re-entry vehicles are the target; kill, as well as intercept, is more difficult. The dense material of the re-entry vehicle and even more dense material of the warhead present smaller, "hardened" targets. They are thus more difficult to track and to destroy. (Causing the nuclear material to explode is not an option. This requires delicate ignitor mechanisms which have almost zero probability of being activated by any type of interceptor-killer. None of the interceptor technologies cause the nuclear warhead to explode. On the contrary, they attempt to destroy it before that can happen.)

These various technological problems of acquisition, intercept, and kill during various phases are not unsolvable. They are complicated and the nature of the problem changes rapidly over the space of a few minutes as the missiles go through the different phases. With a variety of sensors, power supplies, a choice of weapons, and a choice of propulsion devices, some sort of kinetic energy killing device can be make somewhat effective for each phase of the nuclear missile's trajectory.

The "numbers game," however, assures that they will never be completely effective. They will be much less effective than a laser beam with its zero time of flight and constant energy output and where the sensor array remains on the battle station. The laser beam need only dwell on the target for sufficient time to make the kill, a shorter time for a thin-walled structure, a longer time for more dense structures. The kinetic energy device, however, before it is sent on its way, must be known to have the appropriate sensor and sufficient killing power for the particular phase and particular target. But kinetic energy weapons can be engineered to kill some rising missiles and some warheads in the midcourse phase, thus taking pressure off the terminal phase. Most importantly for the argument of this book, though, every device that has been discussed for killing missiles can be used to counter a range of similar type devices based on the satellites.

The problem of killing weapons systems based on a satellite in a fixed orbit is infinitely easier than killing a rising nuclear missile. The time constraints for each of the rapidly changing phases of the problem is relaxed significantly for objects in orbit. The killer satellites are much more vulnerable than their victims, the missiles.

In spite of the more mature technology of kinetic energy devices and their seeming more conventional approach to missile killing, there is one problem which they share equally with lasers and, in fact, any boost phase device. They, like lasers, are designed to begin the battle within a few seconds after the offending missile launches. Decision time in which to react to an alarm of a launching missile is therefore a few seconds also. In this fact we have the primary argument against all launch phase weapons. We will discuss this problem at greater length in section 4.5.

3.6—COMPUTERS, SENSORS, AND BATTLE MANAGEMENT

The most dangerous type of war is when the warrior need not look into the eyes of his victim.

— Dr. Edgar Mitchell

Perhaps the thorniest problem among the new technologies contemplated is that of computerized battle management for coordinating the sensors and weapons for the three phases of ballistic missile defense (BMD). This problem lies at the heart of making SDI work regardless of the weapons technologies employed. The issues are sufficiently complex as to require breakthroughs in computing technology. Simple extension of existing hardware and software technology will not be sufficient for an automated space-based defense system.

As noted in a foregoing section there are currently 1400 missiles in the arsenal of each superpower which must be countered by the other. There are three different basic types of delivery modes: ground launched, submarine launched, and aircraft launched (cruise missiles). They each pose different types of problems. Some of these missiles can accommodate a number of Multiple Independently targetable Reentry Vehicles (MIRVs). The warheads of all types currently number approximately 10,000 for each side, with the numbers increasing each year. In addition, some missiles will carry 10 or more decoys with which to confuse the enemy defense. To approach any reasonable level of defense, all of the weapons in the opponent's arsenal must be countered at once. The logic of this is quite simple. If by launching just one more missile one opponent knows he could saturate the other's defenses, then he would be foolish not to launch the proper number up to the limit of his arsenal. With approximately 10,000 warheads, called reentry vehicles (RV), and at least 10 decoys each, that is a potential of 110,000 targets reentering the atmosphere that must be monitored, tracked, judged as RV or decoy and 10,000 to be fired upon by the defender if they escape launch phase destruction. By making decoy and RV trajectories better simulate each other or by increasing the number of decoys the problem is easily made even more difficult.(Decoys, normally being of different weight and composition, have different trajectories in the atmosphere and different radiation "signatures.")

Naturally any method to reduce the scope of that problem would be welcome. The decoys and RV's blossom out from the missiles during the coasting midcourse phase. The acquisition and tracking problem becomes at least an order of magnitude more complicated at the point of launch vehicle burn out and another order of magnitude more difficult when decoys are deployed. The times involved in the problem are: 100 to 300 seconds for the boost phase, about 1200–1500 seconds for midcourse and 100 seconds for terminal phase, in all no more than about 30 minutes to complete the task of destroying 10,000 or more nuclear warheads among 100,000 decoys launched on 1400 missiles.

During the boost phase, with today's numbers, the maximum problem would be 1400 missiles to fire upon. The detection problem is easiest during launch because the missiles have their hot exhaust plumes which will show once they rise above any cloud cover. Infrared wave-length detectors, even several thousand miles away in space, can detect and lock-on such targets during the boost phase. Boost phase exists for about 300 seconds for older liquid rockets but is decreased with newer, solid, quick-burn rockets to about 120 seconds.[2,15,16] Cloud cover might reduce observable burn time by up to 30 seconds. The burn time can be decreased even further in the future through new rocket technology and thereby further complicate the problem. Let's assume the rising missile starts to become vulnerable to space based lasers after it has risen for about 30 seconds into the atmosphere. Because they are slower, for projectile interceptors the rise time would need to be nearer to 60 seconds before intercept is possible. Thus during boost phase there is a maximum of 270 seconds for older missiles and 90 seconds for newer missiles in which lock-on and killing by lasers can take place; a shorter period for projectiles. How much time is needed for each missile to be penetrated by a laser depends upon the power of the beam, the thermal "hardness" of the target and if it is spinning around its axis of thrust. (Turning on its axis like a barbecue is a defensive maneuver against lasers that can be used to distribute the heat. This increases burn-through time and complicates the defender's problem). Burn through times are estimated to be of the order of from a few tenths of a second to a few seconds for the lasers under consideration.

If projectiles rather than lasers are used to kill missiles, the problem is a classic intercept problem from a distance of 500–1000 miles. If heat seeking guidance is used on the intercept missile, intercept must take place before the aggressor rocket burns out, but not so deep in the atmosphere that the infrared sensor of the interceptor missile is blinded by its own heat of entry into the atmosphere. Thus there is a window of no more than 40 seconds during which intercept could take place. If the intercept is unguided, then it must rely upon explosive "shrapnel" or "bird shot" techniques in the vicinity of the target. The target is likely performing evasive maneuvers during launch to avoid such devices. In this case "homing" guidance is required for the interceptor.

Regardless of which kill weapon is used, however, the computational and communication tasks of battle management are very complex. Coordinating chemical lasers, X-ray lasers, excimer lasers and/or kinetic energy devices in a three phased defense system with only 30 minutes to complete the entire mission faces a formidable, perhaps insoluble problem. The battle management task does not vary significantly with the type of weapon mix that is used except that kinetic energy devices have a slower speed and a more complex intercept calculation. This complexity is insignificant compared with the remainder of the tasks, however. The kinetic interceptor does slow down damage assessment by the amount of the flight time of the interceptor projectile. When the overall engagement time is only 30 minutes, the few seconds required for damage assessment before moving to the next target becomes quite significant. The time constraints in this regard are the most difficult for kinetic energy weapons.

Software experts who have studied the SDI problem estimate that from several tens of millions lines of computer code up to a billion lines of code would be required in order to manage the entire problem.[15,16] It depends upon which software strategies are used as to how much code needs to be written. The least number of lines is still awesome in terms of the number of flawless calculations that must be completed to solve the entire battle problem in a 30-minute span of time with sensors, weapons, battle stations and targets spanning half the circumference of the globe. It is a fascinating technical problem, the consequences: a frightening human problem. No one is certain just how many errors in code it would take propagating through the computing system to totally cause the system to fail. But even small systems cannot be written flawlessly, and a program, or series of programs, of such magnitude as SDI will require many man-years of "debugging" before even a minimum amount of confidence could be placed in there. Certainly such a system should never be trusted to automatically make "go-no go" decisions about the commencement of a nuclear attack.

From our simplistic situation of the warring neighbors discussed on previous pages we discovered that just the fact of introducing automatic machinery into the situation biased the probable outcome toward lose-lose. No consideration of the reliability of the automatic catapults was mentioned. Likely the same result will be observed in this battle management situation. The amount of bias toward lose-lose will be determined by the failure modes and the mechanical reliabilities of the system. One would suspect it would be considerable.

3.7—A SHORT CALCULATION

The most dangerous of our calculations are those we call illusions.
— Georges Bernanos, *Dialogue des Carmelites*, 1949

We will use a continuous wave laser battle station configuration to estimate the number of orbiting battle stations required to fire on all rising missiles during their boost phase. The number of stations is not sensitive to the type laser weapon employed whether ground based or space based. Each will require the same number of killer mirrors and the ground-based system will require additional focusing mirrors in geosynchronous orbit. The number of kinetic energy battle stations will not be markedly different from the number of laser stations. Using for our calculations, first a firing time between targets of 5 seconds, then alternatively 2 seconds, gives reasonable times for computing the scope of the problem. With 5 seconds illumination required, each laser station could attack 55 missile targets which have long burn time (270 seconds vulnerability) but only 18 quick burn missiles (90 seconds vulnerability) during their respective 1aunches. This assumes, of course, that the aggressor conveniently bunches his launch sites so that the requisite number of targets is within range of the laser at the time of launch so that continuous firing is possible.

With 1400 missiles rising it would require 28 laser stations firing at once in the one case and 77 in the other. All would need to be in position overhead or within range of the launch sites, probably about 1000 kilometers from the targets. With illumination reduced to 2 seconds per missile the numbers become respectively 135 long burn time targets per laser station and 45 quick burn time targets per station that could be attacked during a simultaneous launch. If 2 seconds illumination is enough for the kill, then only 11 laser stations are needed in the one case but 33 in the case where all are quick burn rockets. Unfortunately, the laser stations must orbit and cannot stay in one place over the target. There must be a number in orbit elsewhere, spaced so that the necessary number are over the target area at any one time. A factor often used suggests around 10 laser stations in orbit somewhere for everyone on station over the target area.

If we consider the maximum and minimum numbers from the foregoing exercise, we get 110 as the minimum number of laser stations required to fire at 1400 missiles with 270 seconds of vulnerability each and 2 seconds illumination. With only 90 seconds of vulnerability and 5 seconds required per shot, the number increases to 770 laser battle stations to do the job. By increasing the power of the laser station so that less than 2 seconds is required per kill, the number of stations required goes down but the reaction speed for the system must then increase. There is probably a practical limit to how fast the stations can move from one target to the next. That limit is imposed by the mechanical reaction time and inertia of the system.

In the example, the 270 seconds of vulnerability applies to the current generation of Soviet SS-18 and SS-19 missiles, their most numerous. The 90 second vulnerability would apply to the more modern and faster burning U.S. MX missile and the Soviet SS-X-24 which is still under test. If burn times decrease, the number of stations goes up correspondingly.

The purpose of the oversimplified computation exercise was to illustrate the huge computational problem that must be done by computers for the sensors, battle stations anal/or lasers in space. The system must be able to sense, lock-on, point the weapon, fire upon, assess damage, then move to the next target automatically. There must be an automatic method of deciding which of the more than 100 stations is in position to fire at which of the 1400–10,000 targets and then to assign the targets, fire, assess damage, then move to the next target. All of this equipment must be orbiting in space, operating automatically via computer with only a "go or no go" communications link back to the battle command center. The "go or no go" presumably must be made by some human within the first 30 seconds of detecting something that has the signature of a ballistic missile plume. This then becomes the most dangerous point.

With the current strategic posture of mutually assured destruction, the flight time of ballistic missiles between continents becomes the critical time. This amounts to about 25–35 minutes from launch until nuclear impacts begin. This is the time to decide if an alarm is a computer error, an accident, a flock of birds, or an actual attack. Retaliation or not is decided within 25–35 minutes. (A number of false alarms in recent history have proceeded many minutes before being diagnosed as not an attack.) In such periods, missiles are readied and aircraft crews start their engines. The planned space-based SDI systems which can start shooting 30 seconds after lift-off of an ICBM of necessity must have decision making time of no more than 30 seconds after an alarm. It must either be done automatically by a computer or by some human. Neither seems acceptable. Unless one is already at war and shoots at any provocation, more than 30 seconds is required to make a decision to begin a nuclear war. And most certainly one doesn't want a computer to make that decision. If firing is to begin within 30 seconds of a missile launch, then decision time automatically and of necessity is 30 seconds.

Aside from the decision-making issue, the lowest official estimates by advocates of the SDI system require 90 orbiting battle stations to handle a boost phase problem of the existing Soviet arsenal. This number has been attacked as being too small by all critics, but even with 90 battle stations the space battle management problem for the computers is something that has never before been accomplished. (Likely by the time this book is published the number will realistically rise to a few hundred.)

From the small sample calculation in the foregoing, the estimated number of stations changes by almost a factor of ten if one but changes by three seconds the illumination time required by the laser between kills and if one assumes quick burn rocket technology instead of the older liquid rockets. And this is just to handle the boost phase part of the SDI system.

Testimony before the Congress by proponents of SDI has scoped the size of the project as equaling 8 or 9 Manhattan Projects or an equivalent number of Apollo Projects, all to be done simultaneously. Modern technology is wonderful, but it is not clear that it is that wonderful or even can be. But most importantly do humans really want it to be that wonderful even if it can be? We are talking about a war of computers where computers of necessity make most of the decisions from their preprogrammed inputs.

The complications increase dramatically as the number of stations and components increase. And so do the costs. If increasing target illumination time by three seconds and decreasing target launch burn time to one third causes a ten-fold increase in the number of battle stations required, then there will be a corresponding tenfold increase in cost of the orbiting stations. But the computational problem for coordination by the computers increases more than ten times. The computational complexity is not linear with the number of components. In other words the defensive system is very, very sensitive to small changes in the nature and characteristics of the target missile it is trying to kill. And the battle management problem increases nonlinearly with the number of components in the defensive system.

Clearly optimizing the system sensitivity against changes in target characteristics is vital for systems costing a few billion dollars each and which when placed in orbit are almost inaccessible for future changes. Such optimization depends upon what countermeasures the opponent is likely to take. It would be strange if the opponent willingly provided this information during the years that it will take to build the system.

More importantly, however, we have already observed the effect that automated technology tends to produce lose-lose outcomes. It increases the probability of a lose-lose occurrence in the MAD environment.

3.8—THE COUNTERMEASURES PROBLEM

Life is just one damned thing after another.
— Attributed to Frank Ward O'Malley

As soon as the planning begins for any new weapons system, its vulnerabilities must be understood as well as its strengths. This is particularly true for modern systems where the costs and complexities are so large. In any system it is the weakest link that will make the system fail and which the opponent will exploit to the fullest in order to defeat the system. In this regard the SDI concept has so many technological vulnerabilities as to cause question of the concept itself. The first and most glaring is the vulnerability of any orbiting object to physical destruction.

There are many extensions of conventional technologies that can be used as space weapons against other vehicles in space. One can ram, shoot, explode and disrupt vehicles in space with almost any conventional substance that has mass and doesn't instantly evaporate. (Water pistols filled with chemicals against laser mirrors, for example.) Vehicles in space are very vulnerable because they are exotic, complicated, fragile and move in fixed orbits. Even chips of paint from previous spacecraft routinely damage the shuttle spacecraft imparting small pits and scratches. And on one occasion a paint chip almost penetrated the windshield in the cockpit. As previously noted, a one-kilogram projectile at orbital speeds would impart sufficient energy to destroy any spacecraft. New synthetic fibrous materials may be possible for some physical protection, but this only slightly reduces physical vulnerability. It does not eliminate it.

The problem for spacecraft is greatly magnified over that of all other modern war machinery. Modern war machines have highly compacted electronics, computers, radars, and displays. Such delicate and vulnerable components are not easily adapted to the rigors of war where explosives, brute force and heavily armored machines are the norm. This equipment must all be considered as better adapted for "peace keeping" and shows of force against a weaker opponent than for actual fighting. The vulnerability of any modern equipment to disruption or destruction gives the edge to the countermeasure. It is always less difficult and less costly to disrupt or destroy equipment than it is to keep it in operation. And this is particularly true if the opponents are about equal in the sophistication of their measure-countermeasure technology. In the case of space-borne equipment which must continuously orbit at high velocity, the foregoing is overwhelmingly true. They cannot be returned to the shop for repairs in the midst of battle and then be put back into action. One good hit and they are finished! One simply cannot think of the space problem is the same terms as conventional Earth-bound problems. Everything is magnified in complexity by orders of magnitude—except destruction, which is orders of magnitude more simple.

This fact of the economics of build vs. destroy must be given serious consideration when thinking about weapons in space. Any machinery in space which threatens an opponent immediately becomes a high priority target and its destruction is the easiest way to solve his problem. One cannot expect any potential aggressor to ignore the vulnerability of space machinery and the ease with which it can be disabled. No amount of new science can overcome the laws of orbital mechanics and the weight penalties necessary to attempt to protect space-born systems. But these are not problems of new science, these are simply the engineering facts attendant to the environment.

The problem of killing a few or a few dozen laser or kinetic energy machines in space is not terribly difficult. The problem of killing a thousand nuclear tipped machines rising into space and all within the span of a few minutes, is terribly difficult. The opponent's problem of shooting down the battle stations in space or their sensor systems is infinitely easier—provided, of course, that he begins shooting before the system is totally in place and can begin to shoot back. If ground-based lasers are used to kill the battle stations, even the foregoing restriction is lifted. The economic differences are a few hundreds of millions to destroy against a few hundred billions to put in place. Paul Nitze, one of our able arms negotiators suggested that SDI should be "cost effective at the margin."[15,16,17] This means it should be less expensive to build than to counter. After four years of examination, this obviously cannot be accomplished. The effort continues nevertheless with policy papers being rewritten by the administration to establish a new economic criterion. It would not seem that cost effectiveness is no longer a criterion for SDI.

Other American authors and even Soviet scientists have pointed out repeatedly from the beginning of the SDI the ease of destroying satellites. All the anti-ballistic missile weapons discussed in the foregoing are automatically even better anti-satellite weapons (ASATs). Space basing of ASATs is not required, although one could certainly do so. The satellites orbit overhead and thus come to the satellite killer; the weapon does not have to be taken to some remote location. Thus, the first task of an aggressor wanting to launch missiles is to kill the missile killers, then launch the missiles. And ground-based lasers and pop-up lasers are ideal for that purpose. They give no warning, no mirrors are required in space, they can kill a satellite in a few seconds. No space-based systems of any kind are required!

The "numbers game" assures it will require several hundred battle stations for SDI. To destroy 500 orbiting SDI battle stations during the period of one orbit (about 90 minutes) would theoretically require only one ground-based laser firing continuously if illumination times were as much as 10 seconds per satellite. Or one could be ultra cautious and build two ground-based lasers or ten or a hundred. Practically, several would be required near missile launch sites. They would still be less difficult to build and less expensive than 500 orbiting battle stations, irrespective of what technology was aboard the battle stations. To kill a rising missile, the energy deposit on the thin skin of the fuel compartment is estimated at about 20 kilojoules per centimeter squared. Orbiting stations would certainly require more energy to reach a vulnerable spot and to totally kill them. The laser, however, can just maintain the beam on the target until sufficient damage is done whether it takes two seconds or two minutes. Kinetic kills could also be done, less easily, but with equal effectiveness.

The foregoing is one of the central points of this book, as it has been for a number of other SDI analysts after studying the space defense problem. With any conceivable arrangement of high technology weapons, the task of killing satellites is infinitely easier than killing nuclear missiles.

Within a few years, powerful ground-based lasers can be operating. Even if they are so crude as to require an area the size of Manhattan and are so inefficient as to require a 1000 pounds of fuel a second to kill a satellite, they would still be cost effective in defeating any SDI technology.

Yet our "champions" persist in proceeding with this plan to put weapons in space. It is inconceivable that they can be effective. In the measure/countermeasure escalation, the side that concentrates on developing anti-satellite ground-based lasers will shortly deny the other side the ability to maintain in orbit any threatening satellite within the horizon of the ground-based laser. Nitze's criterion of "cost effective at the margin" is an impossible task for SDI, but a realistic criterion. It alone should have killed SDI almost immediately. So far it has not; the Administration has simply changed the economic criteria in its determination to get weapons into space.

CHAPTER 4—ODDS ARE THE HENS LOSE

4.1—THE END POINT

A bad beginning makes a bad ending …

— Euripides, Aeolus

For the sake of discussion, let's begin by assuming that by some means the long-range objective of having a space-based defense is achieved. We will assume it is not a perfect defense but reliable enough to meet current thinking in the range of 90%–95% effectiveness. How reliable (unless the illusive 100% effective) is not crucial to the argument.

The arsenal of necessity would include a mix of weapons on the ground and in space, including lasers ground and/or space based. A number of the technologies previously discussed have been brought to operational readiness. Let's assume that there are a few hundred "defensive" battle stations orbiting the Earth, some with lasers. Some laser stations might be just mirrors tied to ground based lasers. Half of these are controlled by the United States with its allies; half are controlled by the Soviet Union with its eastern bloc. Parity exists.

We must assume that if SDI is successfully deployed by the West that the Soviet could deploy its version as well should they choose. (Do not make the mistake of believing that their technology could not accomplish the task in due course.) Mutual deployment cannot be prevented short of war. Either both sides concur in putting up the defensive stations and do it by agreement or one side begins shooting the other's defenses before they are operational and as they orbit over the opponent's land. If space defense is to be successfully deployed, it can only be so if the opponent allows it to be placed in orbit over his territory. The foregoing would not necessarily be true if all the thousands of tons of components could be placed in orbit simultaneously and were instantly ready for action the moment they were in space. But that is a physical impossibility. All elements cannot be deployed at once. It will require numerous launchings with huge, as yet undesigned, launchers to place the system in space. The first launched must orbit and wait for the remainder of the system in order to be complete, probably many months or even years. Thus the elements are vulnerable to enemy action as soon as they go into space one by one. Placing all the elements in orbit without creating war is going to be difficult. I will discuss that issue subsequently, but for the current argument let us assume that they are successfully placed in orbit.

The most idealistic future scenario would be one with only "defensive" stations in orbit to preserve the peace. All nuclear tipped missiles have been eliminated. President Reagan has suggested this but then reiterated the need for nuclear deterrence and the plan for unilateral deployment if necessary. No subsequent announcements have alluded to the elimination of all nuclear weapons. All have subsequently talked of SDI as an enhancement to nuclear deterrence or as a shield. No one believes at this point that the missiles can all depart. (I will demonstrate the reason in the following.) Thus the realistic scenario has missiles in both silos and submarines with defensive battle stations overhead. Laser stations are any or all of those currently being developed. They can reach down to at least 50,000 feet above the surface and destroy a target the size of an ICBM by vaporizing major parts of it and causing its fuel to explode. There are likely kinetic energy kill stations as well which take a little longer to find the target and are a little less efficient, but which can kill missiles and do little damage on the ground.

The maximum warning time for ICBM's is approximately 30 minutes, less for submarines (about 8–10 minutes from a submerged position) and 4 minutes with the medium and short-range missiles facing each other in Europe. Warning time, although short, is not zero. Total surprise is not possible.

With a laser weapon on the surface or orbiting overhead, the warning time for its target does become zero. Total surprise is now possible.

A laser weapon designed to kill a missile target in less than 2 seconds at 50,000 feet altitude does not cease to be effective at 49,999 feet. How far it penetrates into the atmosphere is dependent upon its wavelength, focus, total energy and the atmospheric density. If the laser light is tuned to windows in the atmosphere, it passes through with some amount of attenuation but sufficient energy to be destructive. It might require 4 seconds to kill an airliner at 35,000 feet and 40 seconds to explode vehicle fuel tanks in Manhattan as well as roast exposed inhabitants in an unbelievable firestorm. And all this without any warning at all! And, with a "defensive" weapon.

Indeed, the 1980s danger of simultaneous annihilation of most of the world's life forms from too many nuclear explosions could almost be eliminated by disposing of all nuclear tipped missiles. But with lasers only, an aggressor wanting to conquer land and resources or just eliminate an adversary can do so with precision and with total surprise. He must first use the laser on the opposing lasers in a simultaneous surprise attack, then he is free to attack anything else at will. Further he can do so without permanently damaging the land and certain other assets he might aspire to control.

Surprise has always been the great fear of defending military commanders and the ultimate strategy of an aggressor! A laser is a weapon capable of instant surprise attacks! But space lasers cannot damage a hardened missile silo, a submerged submarine, nor likely any hardened ground-based object. These assets are safe from lasers or any other weapon except a nuclear warhead.

This is the reason that the "champions" of both sides won't be able to give up their missiles. They feel the need to threaten the other side with total nuclear annihilation in order to feel more certain that "defensive" lasers wouldn't be used for an offensive surprise attack. First strike nuclear missiles are the only weapon that places the opposing champion at significant risk along with his population and serve as a deterrent to a laser attack.

With both sides armed only with orbiting laser stations designed for "defensive" purposes, the superpowers would find themselves back in a position of stalemate and deterrence by threat of retaliation but this time with complete surprise possible instead of 10–30 minutes warning. In the laser station only scenario, the population is completely vulnerable and the "champions" completely safe, since even before SDI, command posts are protected by tons of concrete which are impervious to the orbiting lasers!

The "defensive" laser weapon can become an "offensive" laser weapon capable of setting on fire major cities and vaporizing light metal structures. Life forms will be burned to death instantly—instant sterilization by fire of large areas of the Earth's surface.

To see what experts are saying in this regard, we will provide some quotes. First, Dr. John D.G. Rather, a laser expert and SDI proponent, vice president of the Kaman Aerospace Corporation:

> "Anything that involves large amounts of energy can be used for good or evil purposes. A system of space battle stations designed to stop a nuclear attack also may have the potential to attack selected targets in space, in the atmosphere or down on the surface of the Earth."

> – Philip M. Boffey, "Dark Side of Star Wars: System Could Also Attack," *The New York Times*, 7 March 1985, p. 1

Next, Theodore A. Postol, former adviser on nuclear weapons to the chief of naval operations and an expert on the implications of firestorms:

> "If you were attempting to set fires with an optical laser that was already sufficiently powerful to attack hardened ICBM boosters, there is no question that such a device could also be used to create mass fires of enormous scale and ferocity—mass urban fires potentially larger and more intense than those created by the great incendiary raids on Hamburg and Dresden in World War II." (Ibid.)

It would seem, therefore, that the laser-only threat is insufficient. If nuclear missiles were eliminated, the lasers have no targets except each other and in aggressive situations against populations. Thus they have no justification for existence except aggression. The population is now at risk of surprise attack but the champions controlling the space weapons are not. Thus, the only realistic future scenario including lasers is one where the lasers threaten the missiles if they are used, the missiles keep the lasers from being used offensively against a population, and both sides have first strike missile capability against each other's retaliatory missiles.

Deterrence will have reached a new, more threatening level of stability with two weapons of mass destruction, but MAD remains the same, except the likelihood of its occurring has increased. (Sounds almost silly, doesn't it?)

Rather than increasing deterrence, the weapons in space, particularly lasers, create a new situation which requires nuclear missiles to deter the use of lasers. The natural escalation would be to increase the number of missiles in order to saturate the laser (or kinetic) defenses and restore the "numbers game" balance.

In the words of game theory outcomes: for the champions, the probabilities of win-win are increased because the space defense will kill significant numbers of each other's first strike missiles before they reach silos and command centers. Their win-lose probabilities are about the same as without space defense because the weapons are equivalent. Their lose-lose probability has decreased.

For the population, unless the space defense is perfect (the numbers game again) their win-win goes down because there are now lasers overhead as well missiles threatening them, their win-lose probabilities are still zero and the lose-lose has increased. Note that the populations' win-win or lose-lose are only determined by whether or not the war begins, just as it was before space weapons. Space weapons have only increased the probability that the war will start. So, the population is more at risk but either or both champions can win even if the war starts; the champions win-win has increased because of the space defense.

There is the famous and hilarious Peter Sellers' film, "The Mouse That Roared," which depicts a tiny, medievally armed country declaring war on twentieth century United States. It wants to be conquered immediately in order to receive financial aid to its economy. The "mouse" of the twenty-first century, when all defense might be by orbiting laser plus nuclear missiles pointed at other missiles, will only need to walk into Manhattan with a suitcase-sized nuclear bomb (medieval suitcase, of course).

The current SDI priority of quickly placing kinetic energy weapons in space can only escalate to include ground-based lasers, then space-based lasers because lasers are the ultimate surprise weapon capable of operating against ground or space targets. Directed energy devices (lasers) will constitute the next escalation after kinetic energy devices. Kinetic energy weapons, themselves however, could be scaled up to be effective against surface targets as well, were lasers to take too long to develop. The argument is that space weapons enhance deterrence. Deterrence is designed to reduce the probability of deliberate attack.

But if an opponent has aggressive intentions, laser weapons can provide a means for ultimate surprise, thus they increase the probability of deliberate attack, not deter it. They must be deterred by existing or additional ICBM's. The same is true even for "offensive" kinetic devices in space, except the warning time rather than being zero is of the order of one or two minutes.

The existing SDI effort can produce only one positive effect, and it can have that one only if the attendant publicity and public debate were to lead to public demand for the beginning of de-escalation of the entire process to a lower more stable level.

This can be accomplished through cooperative effort and arms control. The process will be much easier to accomplish now than later after space deployment has begun—just begin with arms control and cooperative effort in space initially if the desire is to make the world more secure. The proper high priority objective should be to reduce the nation's lose probability rather than that of the military!

There is no responsible person who would suggest that deterrence against overt aggression, against petty tyrants and terrorists, against rogue dictators, against the ever-present criminal element does not need to be maintained. A weapons-free and conflict-free society of humans is a dangerous illusion. Thus the question is not reasonably one of how to eliminate weapons, but of how to reduce the reliance upon weapons of mass destruction as a means of managing major conflict. Both nuclear warheads and orbiting lasers are weapons of mass destruction. The decision must be that of choosing the proper precision weapon instead of the massive destruction of nuclear energy or even of laser firestorms. Now that nuclear knowledge is abroad, it can never be recalled.

But a terrorist with a suitcase nuclear bomb can be controlled with a big stick or any number of weapons short of another nuclear weapon if he can be controlled at all. Should nuclear weapons ever be eliminated we must live with the fact that someone, somewhere, sometime can create another nuclear arsenal covertly. But they are not likely to covertly create a significant arsenal if civilized nations cooperate in controlling nuclear materials. Fear of a rogue nation amassing a few nuclear weapons covertly does not constitute reason for civilized humans to perpetuate their own massive arsenal as a deterrent. It would not be.

Let us address one more topic in this idealized end-point scenario of weapons in space. Were it to be accomplished as in the foregoing, each side has not a "shield" to protect itself but a "lid" over the other country's. This means that each side must at least advise, but more likely get permission of the other in order to carry on its normal and routine operations at high altitude or in space. Not to do so risks having routine equipment deliberately or accidentally destroyed without warning. It is impossible to imagine the United States permitting this condition to exist. But neither is it imaginable that the Soviet Union will permit us to likewise threaten their sovereignty. To attempt to unilaterally create this condition can only cause the shooting to commence. No sovereign nation can permit its routine functions to be made subject to the veto of a hostile power if it has the means to prevent it.

4.2 —THE REAL SITUATION

See how today's achievement is only tomorrow's confusion;
See how possession always cheapens the thing that was precious.

— William Dean Howells, *Pordenone, IV*

In the previous section we began by postulating the end point of a laser "defensive" system held by both the U.S. and the Soviet Union in addition to an arsenal of nuclear missiles. We discovered that such an end point would leave the adversaries in a more precarious position than the one without space weapons. The end point of lasers and missiles presumably could be viewed as a stable condition if both sides are again in balance with similar weaponry, should we be able to reach that point without beginning a war.

The Soviets brought their nuclear capability from behind in the 1960s up to a point of relative parity in the 70s. The condition of the 1970s and early 80s was a stable point with only nuclear missile deterrent, without significant first strike capability. Modernization and upgrade toward first strike capability by both began to destabilize the confrontation in the 1980s. Then SDI was announced. The time interval between the '80s and the next stable condition is likely a period of several decades in which the new technologies are engineered, tested, and made operational, with first strike capability being more closely approached on both sides. This period can only be assessed as highly unstable and dangerous. The most likely time for aggressor action, for preemptive strike "defensive" action or an irrational event is during this unsettled time of an all-out race to reach a new balanced and "secure" defensive posture. We will examine the transitional possibilities between these presumably stable conditions. But we will also point out the impact of reduced warning time and first strike capability on the stable condition, if in fact it could be reached.

Today we have an arsenal of about 1400 missiles on each side and about 10,000 warheads, each which is increasing slightly each year if Strategic Arms Limitation Talks (SALT) constraints are maintained. It will increase faster without SALT constraints. Progress toward obtaining significant first strike capability on both sides is underway because of modernization and upgraded technology. Fortunately, replacing or retrofitting 1400 missiles cannot be done quickly or the instability would even be greater. This modernization is, in the 1980s, a fact of existence that is not likely to be reversed and thus must be endured to its end point.

The scenario of the middle period is not difficult to construct. It corresponds to attempting to build everything at once, in the way of weaponry, for which humankind has the least need, limited only by economic restraints. But it makes the champions happy and keeps the factories on both sides busy producing the pieces. The existing ABM Treaty and SALT accords are likely abrogated since the activity is inconsistent with either. We have the existing capability for massive nuclear retaliation with the warheads on each side. We have increasing numbers of missiles, both cruise missiles and ballistic missiles, with targeting accuracies such that hardened silos and command centers are vulnerable. We have a variety of kinetic energy and laser battle stations going into operation which can kill some of the ballistic missiles, but as yet none of the cruise missiles because they fly too low. Airborne or space radars must be developed to find and track cruise missiles. But by increasing laser power and deploying new phased-array radars for detection, ground, airborne and orbited, even cruise missiles and ground targets can become vulnerable. This or some other cruise missile defense must be developed under the Air Defense Initiative.

But because each does not have perfect and reliable inspection of all the opposing champion's satellite payloads, it is necessary quickly to have satellite killers to shoot down rogue satellites in case someone's champion slips a payload with "offensive" capability into orbit. In fact, because the planned "defensive" battle stations have significant offensive capability, the ASATs are already necessary in the minds of planners. They are necessary also to "blind" observation satellites in case attack becomes imminent. The ASATs must have the capability to reach high orbits as well as low orbits. Low orbit satellites are relatively easy to kill, higher orbits require a bit more sophisticated technology. But with a powerful ground-based laser, any threatening satellite can be killed by sufficiently increasing the dwell time of the laser beam on the target. Alternatively, explosive space mines in orbit or kinetic energy counter measures may be used against satellites.

In this book we have not emphasized the problems of mid-course phase and terminal phase defenses because it is the search for boost phase defense that sets the whole escalatory system in motion. But because the boost phase weapons can likely not kill more than 70–80% of the launched missiles, the mid-course and terminal phase defense must be able to screen out the remaining nuclear reentry vehicles from among decoys, chaff, and other screening devices. Each phase of the tracking and killing of missiles has its own unique problems and requires special equipment for the job in order to reach the total effectiveness desired. The problem of tracking the "cold" targets in space requires new sensors, likely new radar on the surface, and perhaps in space as well. Big radars require too much energy for space use unless nuclear power is used. Thus nuclear power technology for space use will need to be expanded to the output needed. Ground based lasers guided by radar represent the best possible long-range solutions to the mid-course problem. Kinetic energy or nuclear devices represent the best solution to the terminal phase. As we know, kinetic energy devices can be used in the mid-course phase while lasers are being perfected. Thus there will be a mix of these weapons. Tracking during these latter phases and distinguishing the decoys is a significant problem, however. The sensor must "look inside" the target and determine if it is inert or an active device. This can be done with the appropriate frequencies, the same as one can get a chest X-ray. But it must be done in the space environment and while moving at high speed. This will require new active sensing devices.

It is possible that for the mid-course phase problem the X-ray laser instead of some other laser would be chosen. It could be a mid-layer "defensive" weapon (unless used against satellites, then it is "offensive"). But with the X-ray laser we have the most dreaded of all weapons, a nuclear bomb in space. Once the treaties are broken and there is one nuclear bomb planned for space, there is not a rationale or a means to deny others. To plan to launch a nuclear-powered X-ray laser from the surface at time of attack just opens the door to escalation so that nuclear explosives orbit in space. To de-orbit a nuclear warhead from space requires a couple of minutes rather than the 10–30 minutes required to shoot a missile up and let it come back down. Warning time is thus decreased and surprise more attainable. Even to entertain seriously the idea of nuclear explosives in space is even more destabilizing than lasers in space. (There is a significant difference between nuclear power plants in space and nuclear explosives in space. Although the former are not very desirable in Earth orbit, they do not represent the huge threat presented by nuclear explosives.)

During the mid years, then, between the 1980s and whenever endless technological escalation could produce a laser defense system, we have the worst of all worlds, much worse than anything humanity is experiencing now in terms of fear and uncertainty. The massive kill capability of the 2800 missiles and 20,000 warheads has not gone away but has increased by 30–100%. The targeting accuracy of all guidance systems has improved to permit pinpoint targeting. Underwater detection will have improved so that submarines are no longer totally invulnerable and thus are more subject to first strike. As significant progress toward first strike capability is obtained by one side, decision time is shortened for the other side. Whereas without first strike targeting, one side could take the hit and then retaliate, the champions must not wait if incoming missiles are known to have capability to kill missiles in the silo or underwater. Retaliation must begin before incoming missiles can land or else the capability to retaliate is lost. This is called a "launch on warning" condition. With both having first strike targeting capability, decision time is less than the warning time for both sides. The effect of reducing decision time is to increase the probability of retaliation by accident or mistake. But notice that first strike targeting capability does not change the balance of power or in any real sense threaten retaliatory capability. The only real net effect is the reduction of decision time for use of missiles against missiles. This in turn increases the likelihood of accident.

Orbiting overhead are kinetic kill weapons being replaced and upgraded to laser stations (or mirrors) which have a warning time of zero for whatever their intended targets. There are not enough of any weapons in the mid years to be maximally effective since it will take considerable time to get enough in place to fire at all ballistic missiles. The population is now threatened from overhead with zero warning time as well as the missiles with 10–30 minutes warning time. The champions have had their nuclear retaliation decision time reduced to less than 30 minutes. Should laser attacks occur on population or high-flying military aircraft, decisions must be made as to the use of nuclear missiles. And the nuclear bombs that might have been concealed as an observation satellite can de-orbit in a couple of minutes.

What if that should happen? Once the ABM Treaty is violated, anything can happen regarding ASATs and weapons in space. Innocent looking defensive lasers can actually be higher powered offensive lasers. The strategic balance is not a balance but an uncontrolled melee during the middle years between the stable points. And that is what is being "sold" as greater stability and as a way to reduce the nuclear threats. It seems highly unlikely that a new stable point can ever be reached before a war actually begins.

4.3—FIRST STRIKE AND RETALIATION

The art of war is simple enough. Find out where your enemy is. Get at him as soon as you can. Strike at him as hard as you can and as often as you can and keep moving on.

— U.S. Grant, *On the Art of War*

General Grant's statement from the 19th century is precisely the type thinking which is now antiquated by 20th century conditions.

Because the idea of "first strike capability" is used ambiguously, sometimes differently by different authors, the term often appears to be the shell in a "shell game." First you see it, then you don't. It sounds like a rather precise phrase but in fact is not. This is not altogether by accident. It permits confusion and uncertainty to reign when budgets and approvals of plans are needed. Uncertainty normally helps in the appropriation process for defense budgets. The error is never to the side of too little. One can hear or read the arguments regarding first strike, then the conclusions, and be totally in doubt as to how they connect with each other. Yet by some means the budgets get approved. It would appear that ambiguous definitions, liberal use of technical acronyms, and a lot of fast talk is indigenous, even vital, to the military budgeting process.

It seems important to touch upon the details of the "first strike" ambiguities in this work in order to understand what is being implied in the public reports, to eliminate confusion and to better understand the SDI problem.

There are a number of definitions of "first strike capability." The one with the widest acceptance by the military is: a strike such that the opponent is disarmed and thereby precluded from retaliating. This notion can be used globally in a strategic sense, or it can be used locally in terms of limited objectives. Unless one specifies the scope of the meaning each time, confusion is certain to arise because vastly different things are implied if the term is used in a strategic sense rather than in a local sense.

The more limited notion implies three technical necessities: 1) that the position of the target is precisely known, 2) that guidance accuracy is sufficient to deliver a nuclear warhead within kill range of the target, 3) that the explosive is enough to damage the target. The latter is not a difficult consideration because nuclear warheads are made in a number of sizes and presumably the appropriate warheads can be mated to the targets. The first item is not a difficult consideration if the target is a fixed installation. Thus it is guidance accuracy that is at the heart of the notion of first strike capability. This is an issue whether we talk in terms of local situations or a global strategic situation. It always has been a limiting factor. In order to damage a target sufficiently to preclude its further use, the circular error probable (CEP) of the guidance system must be small enough to deliver the appropriate warhead on target. In order to kill hardened missile silos, CEPs of no more than a few hundred feet are required.

Neither side has been able to do that at intercontinental distances until the last few years. But improvements in guidance technology have always been on the side of the U.S. The hardening of underground targets has been a Soviet priority.[2] The MX and Trident II missiles have CEPs of the order of a few hundred feet at appropriate ranges. The SS-18, the most predominate Soviet ICBM has a CEP of 1200 feet, which is too large to be considered as first strike capability against hardened silos. From the guidance point of view, the U.S. has better first strike capability even though later version Soviet missiles will all have improved guidance, and they are improving the "hardness" of their silos.

In the strategic sense of "first strike capability," that is, to disarm an enemy completely, we must also add a fourth item to the foregoing three requirements, It is: have sufficient weapons with accurate guidance to engage ALL the opponent's retaliatory capability. It is this last item that drives the escalation of the "numbers game" of the strategic forces. Each seeks to have too many weapons for the other to destroy and thereby be invulnerable to being disarmed by a first strike from the other side. The addition of improved targeting capability permits the attacking side to use fewer weapons to neutralize the other and requires each "defender" to have additional sites or more hardened sites in order to avoid being totally disarmed. From this viewpoint, "defensive" necessity can be said to drive the escalation rather than "offensive" intent.

With regard to the targeting problem associated with first strike capability, mobile targets or concealed targets introduce uncertainty into targeting. Submarines provide both and thus are the most invulnerable to attack by ICBMs. The U.S. has far more of its capability in submarines than does the opponent.

We find, therefore, that within the notion of "first strike capability" we have embodied the following significant concepts: 1) guidance accuracy sufficient to deliver a killing blow on an installation; 2) mobility of retaliatory targets so that their location is uncertain; 3) the use of first strike to obtain limited local objectives; 4) the use of first strike to obtain total disarming of an opponent in a strategic sense. And with this latter sense, 5) sufficient weapons with the necessary capabilities to accomplish item number four. Thus, depending upon with whom one is talking, any or all of the five items may be at issue in the same discussion.

First strike seems to be a precise notion but is in fact a "catch-all phrase" containing technical, policy, tactical, and strategic implications. Military leadership (likely on both sides) routinely use the phrase to imply "they" have it and "we" don't, then the advocate bounces all around the various aspects of the phrase without being very specific about what it is "they" have that "we" don't. We will see in the following that the correct answer is "nothing."

In terms of guidance capability, which is at the heart of the matter, the U.S. technology has consistently been in advance of that of the Soviet even though that lead is thought to be decreasing. In terms of the strategic sense of the phrase, which implies sufficient numbers of weapons to totally disarm the opponent, the Soviet arsenal only began to catch up with the U.S. arsenal during the 1970s and with a weapons mix considerably inferior to the U.S. with respect to guidance capability. And, with 60% of U.S. strategic missiles mobile and concealed under water in submarines, the Soviet ability to achieve a strategic "first strike" has zero probability even though their guidance systems are improved.

By whatever measuring stick is used: targeting, number of weapons or location of weapons, the distance to go between existing capability and having a first strike capability has always been shorter for the U.S. intention to attain the goal of having first strike capability is really not the issue. The modernization of strategic systems coupled with the relentless growth in the number of weapons assures that both sides approach ever closer to that condition whether or not it is an avowed goal. "Defensive" necessity and modernization is the rationale on both sides. It is finally the invulnerability of submarines that precludes strategic first strike from being a near-term possibility for either side.

But with the Soviet having less than 10% of its capability underwater and the U.S. nearly 60% it is not difficult to ascertain which is the most invulnerable to deliberate first strike. In addition, U.S. capability of deep offshore detection of submarines is ahead of Soviet technology in this regard so U.S. underwater retaliatory capability is not only greater in number it is more secure. It is foolish to deny that this situation exists. It destroys official credibility. (It is self-contradictory for the military to claim superior guidance technology and a greater percentage of submarine-based missiles, then claim that the Soviet has more complete first strike capability.)

These facts naturally create more concern for the Soviet than for the U.S. They are why the Soviet is trying to increase its submarine fleet, improve its guidance and further harden its silos. These discrepancies in the details of specific capabilities are not a major problem, however, so long as there is sufficient retaliatory capability to make an offensive strike from either side unrealistic. Rhetoric which implies otherwise sounds like the small shepherd boy crying "wolf." The military has little to do in peace time but keep its weapons ready and worry that the other side is getting better. Since 1980, the requests for more budgets sound like: "Wolf!"

In the foregoing, we touched upon the idea of mobile targets making first strike difficult. Uncertainty in a target's location does deter the opponent from shooting at it. That is why submarines are so useful. Uncertainty in targeting is also the concept which drives the desire to make land-based missiles more hard or more mobile, or to have more silos than missiles so that the opponent does not know which one to point a missile toward. There have been numerous schemes to accomplish this, all expensive and none which can provide the targeting uncertainty of a submarine. The valid questions on this subject are: When is enough, enough? How many moving targets are required to make the opponent certain that he will get more back than he sends? Isn't 60% enough? Or is it inter-service rivalry for more money that is driving this idea?

Complicating the opponent's targeting problem is a legitimate tactic, peace or war, new rules or old rules of the game. But how far is it necessary to go in this regard? Even though the military does not like to acknowledge the notion, it remains the mutually assured destruction of the populations and national assets that ultimately prevents strategic nuclear weapons from being used deliberately in an offensive mode. That is the real deterrence. It is not, in the final analysis, the uncertainty as to where the missiles are that is the key to deterrence!

To illustrate: if the location of all targets were perfectly known and all warheads had a CEP of zero feet, it is the fact that a "launch on warning" policy would still assure that massive retaliation on each side could yet be accomplished to destroy national assets. Retaliation would be against population since the "offensive" missiles have already been fired and the offending silos are empty!

Thus a deliberate, i.e., rational, first strike would not occur (with rational leadership) because it is certain that their own population would be destroyed. The real deterrent is ultimately the threat of destruction of one's own nation. But this sort of reasoning forces recognition that so long as even minimal retaliatory capability exists to destroy the population and national assets of both sides, deliberate nuclear war will not take place. Only minimal retaliation is necessary, not the maximum. That admission would ruin the "budget game." The military of both sides enjoy their traditional game of "one-upmanship" and increasing budgets, while preferring to believe that the population's risks and outcomes are the same as their own, when in fact they are not. The technological breakthroughs of the past 40 years have changed that reality. The policy of deterrence and projects to enhance deterrence center on deterring a deliberate first strike occurring from the other side. But it is not the deliberate strike that is the danger in the modern era. It is the accidental or irrational strike. Deterrence does not address that issue. Recognizing that the probability is increased MAD from accidental causes would properly deal with the issue.

From the targeting problem discussed in the foregoing, uncertainty in target location is seen to diminish the ability to achieve a first strike. Proponents of SDI generalize this notion and suggest that any uncertainty in the opponent's mind about his success enhances one's security. By disposing of a significant number of attacking missiles, SDI would presumably create greater uncertainty for the success of an attack thus it enhances deterrence. If deliberate aggressive attack were the real concern the argument might have some validity. But because the probability of accidental attack is increased thereby, it does not.

I will argue in the section "Better Solutions" that, in general, creating uncertainty for the opponent is not the road for reducing tensions nor to solving the MAD problem. Only in the limited sense of creating uncertainty in target location can I agree, because this does help toward defeat of first strike capability without increasing the likelihood of an accident. Even though first strike is not pivotal, as pointed out, it is important in that risk of accidental war increases as both sides gain the technology of first strike targeting. Thus, creating uncertainty of target location through mobile missiles enhances deterrence against first strike and is stabilizing. That is not true for all uncertainties which can be created, and each side must be careful not to use the broad notion of "creating uncertainty" as a generally valid concept. I will demonstrate in section 5.4 that it is not.

We need to address one more topic under "first strike" before leaving this subject. There exist within U.S. strategic planning certain options for using nuclear missiles on a limited basis to achieve limited military objectives. Likely, similar plans exist in the opponent's planning rooms. In other words, there is still belief within military circles of a limited nuclear war against our superpower opponent. No one else believes this is possible, and the odds against containment once nuclear exchange begins are high, indeed too high to require much discussion. It is difficult to believe that in this era such plans still exist. With all the public rhetoric proclaiming peaceful intentions; with all the enormous risk that any nuclear exchange between the superpowers will inevitably escalate to total nuclear exchange; with the "alarmed" declarations that the opposition is gaining first strike capability, how can such plans and such thinking still be with us?

There is no useful purpose conceivable that such plans can serve. They only emphasize the fact that military policy is still tied to traditional win-lose strategies.

Perhaps they explain the preoccupation with "deterrence of deliberate attack" when in fact deliberate attack is not the real danger. They serve to emphasize that MAD is not acknowledged by the champions as the necessary outcome of their activity, should any active fighting begin between superpowers.

4.4—WARNING TIMES AND FALSE ALARMS

*Thus we play the fools with time and the spirits of the wise
sit in the clouds and mock us.*

— Shakespeare, *Henry IV, part II*

In prior sections of this book the issue of warning time and decision time has been emphasized. This is such a crucial question for the future of the world that it must be explored exhaustively before future systems should be produced. In the current posture of deterrence, both sides have learned to accommodate warning times of no more then 25–30 minutes. Without significant first strike guidance available for either side, decision time can be greater than warning time because a strike can be sustained, and retaliation still take place. In the case of missiles stationed in Europe and with launches from the submarines of both superpowers, the warning time decreases to under 10 minutes. But as long as warning time is finite, the net effect of first strike capability is to reduce decision time to slightly less than warning time. First strike capability does not preclude retaliation. The militaries make considerable point of the other side gaining first strike capability. But as long as warning time is non-zero, only decision time is really affected.

Warning times of 25 minutes or even 10 minutes, while uncomfortably short and dangerous, do leave some room for error. False alarms, accidents, and computer failure can be (and have been) correctly diagnosed before retaliation is begun. The real deterrent has been the assured massive retaliation itself if an actual strike were to be made. The probability of a deliberate launch is zero or very near zero under these conditions even if the opponent has aggressive intentions in other ways. Thus the probability of a false alarm is greater than the probability of deliberate attack. (The proof: there have been many false alarms and no deliberate attacks during the years of this condition.) The political situation may be uneasy and fretful, but it is relatively stable because the probability of deliberate attack is so low. When a warning is actually received, however, the situation changes. After the alarm is sounded it must be treated as real until diagnosed as false. At such time the probabilities do not matter. Either the false alarm or the attack must be confirmed absolutely. The defender must behave as though it is an actual attack until he can prove to himself that it is not. If the opponent does not have first strike capability, the defender can even allow the warheads to explode in order to be absolutely sure, should reasons for a false alarm not have been discovered.

This stable situation is no longer the case as each side uses new technology to increase the accuracy of its targeting information and of its warhead guidance and control systems. Whereas missing hardened missile silos and control centers by 1000 feet or more cannot assure their destruction, hitting within a few hundred feet with two missiles can. The question of warning time now becomes critically more important. The more certain is the other side's first strike capability and the shorter the warning time, the more likely is a retaliatory launch to be ordered before incoming missiles can land (a launch on warning condition) and the more likely a preemptive first strike from desperation. (We will henceforth call such a desperation strike an irrational strike.) Likewise, it is more likely that an error, accident or computer failure cannot be detected before a false retaliatory strike is begun when decision time is shorter than warning time. (We will henceforth call such a strike an accidental strike.)

Even though the probability of a deliberate attack remains near zero, the introduction of first strike capability causes the probabilities of irrational attack and accidental attack to increase. How much they increase is a function of decision time, diagnostic time and the nuclear response decision making process of both sides. If the system diagnostic time (the time required to test the system for failures after an alarm occurs) becomes identical to or exceeds the warning time an exceedingly dangerous condition comes into being. This is particularly true for the Soviet side where computer systems technologies are markedly inferior to those of the United States. Were the U.S. to attain complete strategic first strike capability the likelihood of an accidental strike from the Soviet side is significantly increased because of their less reliable computer technology. Such an accidental launch would likely be interpreted on the U.S. side as a preemptive first strike.

Notice there is no talk of cities and people in this discussion, just killing missiles and silos and other military objectives. People and cities can't be considered once the alert is sounded, be it real or a false alarm. It is too late then. People must be considered in the a priori analyses before systems are deployed. This is the reason in the simple analogies that the outcomes for population and the champions were considered separately. It is the same in the actual situation. Once the automated technology is introduced, it is the functioning of the technologies and their response times that drive the system after the alert takes place. This causes people's risk and the military risk to be considered separately. The people and national assets may already have been lost from an actual strike and the champions are still engaged in response and retaliation through the automatic systems.

The Soviet retaliatory capability is currently more in jeopardy than the U.S. Therefore the prior likelihood of an irrational first strike or a false alarm accidental strike coming from the Soviet side has increased. Even though first strike capability in the strategic sense may not be complete on either side, at some point any alarm will be treated as though the presumed incoming warhead had first strike capability. One cannot know, in the event of a real attack, which warheads do and which ones do not have enhanced guidance. Thus after a certain number of missiles are known to have first strike capability, the effect on the other side is to treat any alarm as though all opposing missiles were so equipped. Thus even though first strike capability may not be complete on either side, the opponent must begin to react as though it were and this increases the likelihood for launching an accidental attack.

Let's deal with the irrational attack scenario. The conditions where this might arise originate in command and control, that is, the decision-making function. We will rule out for the moment the possibility of madmen gaining control of the machinery. That is conceivable on either side but improbable. Both sides have elaborate decision-making systems to prevent such an occurrence. The irrational attack to be considered here is faulty assessment and decision making because of decision times becoming too short for the decision-making system to function adequately. This can lead to a preemptive strike when in fact a strike from the other side was not forthcoming. With existing conditions this has not been a significant probability. Attaining first strike capability would cause the irrational attack probability to increase to significance should "launch on warning" be used by either side as a response to first strike capability on the other. The probability of an irrational attack decision becomes large if the time to reach and communicate a decision is longer than the warning time. It is analogous to diagnostic time in this regard. Irrational attack has to do with the human decision-making process and the length of time required to make rational decisions. Accidental attack is determined by the automated system reliability and the time required to diagnose a false alarm. When either or both of these processes are as long as warning time, the probability of an erroneous launch becomes large.

The warning time problem becomes even worse when we put weapons (presumably defensive ones) in space. We have already established that the spaced based systems for boost phase defense can commence firing when an attacking missile is about 30 seconds into its launch. In fact, they must begin firing within about 30 seconds if boost phase defense is to be effective. Decision time is now automatically 30 seconds. As soon as a launch is detected, i.e., the alarm is given, the clock begins to run on decision time and diagnostic time. And all the foregoing scenario on diagnostics and decision making is compressed into a 30 second time span instead of 30 minutes, almost two orders of magnitude shorter. This seems to be an unrealistically short time for humans or computers to make a decision to fire on another country's missiles unless the war has already begun. It is an impossible peace time scenario for making a decision to go to war. How does one launch a perfectly normal and peaceful space mission in such a case. If the system is fully automatic and computerized (as is being considered), the space defenses must be turned off to allow the peaceful missile to be launched. But what if the peaceful mission is concealment for an actual attack?

And what if a laser is fired in space? If it is pointed at the surface, something is going to get burned instantly. Was it an accident, a false alarm, a computer error or an attack? There is no time to warn the other side that an accident or computer failure took place. If it were an accident or computer error, it would likely set off the other side's automatic response system, nevertheless. The weapons in space scenario presents dozens of troubling situations of this type just because decision times approach zero. With any system of computerized battle management, a software error or a small hardware failure is likely to propagate through the system in unknown and unpredictable ways. With an instantaneous weapon of mass destruction, a laser, at the end of the chain of command, a system is clearly in place for accidental stimulation of war. But even with kinetic energy systems the decision problem is not relieved. Decision time is still only 30 seconds. A system designed for boost phase targeting must of necessity have a decision time and diagnostic time of the order of 30 seconds. It is a far higher risk than has ever before been managed. It is most doubtful that humans or computers can cope with this limited time scale reasonably.

But consider the end point problem again from this point of view. With lasers in space and zero warning time, any strike is a first strike, regardless of whether it is accidental or deliberate and regardless of whether against ground targets or orbiting targets. If a crisis develops between the opponents such that forces are brought to the maximum state of readiness, any perturbation to the system that produces an alarm likely will be considered as an actual attack. The beginning of retaliatory action must instantly start to avoid orbiting defenses being destroyed. The existence of zero or near zero warning time weapons requires that decision making and diagnostic procedures be of the same order. Were this not so, the opponent could in an instantaneous surprise attack destroy all the missile defenses. This seems to be a totally untenable situation in which humans on both sides might find themselves. The options are to use the system and risk an accident or discard it unless it is 100% reliable. But there is no such thing as a 100% reliable machine. And the risk seems unacceptable.

4.5—RATIONALIZATIONS

The rationalization of desires rather than desire for the rational is a source of untold human conflict and strife.

— Dr. Edgar Mitchell

There have been three different reasons put forth by advocates as to why the Strategic Defense Initiative should proceed. They are: defensive necessity, a bargaining chip, and economic spin-off for arms reduction. In light of the arguments that have been presented on these pages, let us now examine each of these justifications in turn.

4.5.1—DEFENSIVE NECESSITY ARGUMENT

Necessity may be the mother of invention, but as often 'tis the argument of the self-serving.

— Dr. Edgar Mitchell

It should be reasonably clear at this point that since humankind has loosed the bonds of gravitation and ventured into space, the "rules of the game" for warfare must be changed dramatically. The problems arise from lack of political awareness of that fact. The legitimate need for defense which has been with us since the beginning of history must be looked at in a new, more discerning way as a result.

Preparedness for attack has been the historic way defenders have viewed the issue. Mobility and surprise has been the fundamental strategy of any aggressor. In this current era, whether defender or aggressor, each has an ultimate weapon for mass destruction—nuclear energy—and is developing another—lasers. While less destructive and "cleaner," the latter are more precise and permit the aggressor's desire for ultimate surprise: zero warning time. Perhaps there is something more deadly than the combination of these two but if so no one can imagine what it would be. We have tied the nuclear weapon to a space borne delivery vehicle which can reach halfway around the world in 30 minutes. With space basing that time reduces to two minutes.

There are only two drawbacks to this situation. The defender has identical weapons, and the firepower is so destructive that it renders conquest by this means useless because it destroys all the assets. Thus the opponents are reduced to stand-off with no possibility of distinguishing between aggressor and defender except by their intentions, which each one knows only for himself. Now add an orbiting laser that can strike anywhere in the world or in space instantly. The defender and aggressor both have two ultimate weapons, one for destruction and one for surprise.

Space provides the means by which weapons can be orbited over any and every point on the Earth, thus providing total mobility and potential threat to everyone. Furthermore, by being aimed selectively and accurately, the laser weapon can destroy first the space defenses and then only those assets of the opponent that the aggressor chooses. Except that the laser cannot destroy the nuclear missiles unless they are launched in an attack. Thus the missiles are the only "defensive" assets safe from a laser attack, as long as they are underground or underwater and not in use.

With such a capability in existence, no sensible "defender" is going to permit a potential aggressor to possess a mobile weapon of total surprise if that can be prevented. But because the warning time is zero when the laser is in space, there are really only two options: destroy them before they can be used or come to terms with the opponent and find ways of policing military-industrial activity that assures the weapons are never built. If the first option is chosen, then the situation that everyone tries to avoid occurs: war between the superpowers. A preemptive strike against any existing weapon of total surprise is the only means of defense. The defender must become the aggressor in his own defense. And because orbiting objects are so vulnerable, destroying them in space is the defender's most logical and economical course of action.

Kinetic energy weapons in space must be viewed as just a step in the direction of placing other weapons in space, either lasers, nuclear weapons, or both. Treaties must be broken to place any weapon in space. What possible rationale could lead one to believe that space weapons could be limited to kinetic energy weapons only, if more powerful weapons can be built? There is just no reason to believe that could happen. Kinetic energy weapons create most of the problem by themselves. Kinetic energy weapons, just as the lasers, attempt to place a "lid" over a sovereign nation in a time of relative peace. They reduce decision time to the order of 30 seconds, the same as the laser. Like lasers, they can preclude a sovereign power from using the upper reaches of atmosphere and the space above its own land without the acquiescence of an opponent. It is clear that no sovereign nation can permit this if the means exists to prevent it.

Obviously, this situation requires a new point of view, new "rules of the game." The role of defenders and aggressors has never in history been so murky and confused.

Each military has a legitimate need to defend but must become aggressors to do so. To use a phrase which has been attributed to Albert Einstein: "A problem can never be solved at the same level from which it is created. One must always go to some different level." In this case a simple reexamination of the problem is sufficient. The problem has always been stated in terms of "defending against attack." But that isn't the true problem. The true problem is: "achieving freedom from the fear of attack." Or if that is asking too much immediately, perhaps: "freedom from the fear of a surprise attack." It is the "fear" that is at the root of the problem and then secondarily the "surprise." Let us address these true issues instead of the one that can only exacerbate the situation.

Because building weapons systems of such magnitude and complexity require a significant amount of time and activity, it is not easily concealed even if one has the technology in hand. Surveillance and inspection can provide freedom from surprise if opponents truly want just to defend themselves. Long lead times and heavy industrial activity are required to mount the new generation of weapons. Undertaking to build this new generation for space can only be construed by an opponent as prima facie evidence of aggressive intentions. Not building them and open inspection to demonstrate the fact, particularly with the capability to do so, can be construed as peaceful activity. Thus to prevent a "shoot it out" solution to weapons in space, there is only one other alternative: agree not to build any of them and provide adequate assurances they are not being built.

The alternatives stop at this point. No others exist. If both agree to orbit them, the fear increases. And it is fear that is the primary root of the defense need. Agreement by opponents to both orbit reduced warning time weaponry may seem like "defense" but it only exacerbates the real problem which is fear of attack. Fear comes from two sources: the first is a valid and demonstrable concern that the opponent has aggressive intentions, that is, he is demonstrably willing to launch a deliberate attack; the other is a paranoid preoccupation with the opponent's presumed aggressive intentions. Solution to the first requires that both opponents demonstrate to the other their peaceful intentions.

This requires fully understanding what the opponent is afraid of, his legitimate defense needs, and then taking steps not to exacerbate his fears. If both are in reality, however, playing the game of power politics and vying for world domination, but calling it something else, then some steps at achieving honesty and good faith discussions must come first.

It may be that the only good result of the SDI exercise is the recognition on both sides that the stakes are too high for posturing and power politics as usual. The fate of the world hangs in the balance and errors in judgment or in assessing the opponent's intentions cannot be tolerated. The situation requires straight talk and honest assessments, otherwise there is little hope to reduce the fear level before shooting begins. "They blinked first" is hardly a mature response to the problem.

Regarding the second cause of fear: the paranoid preoccupation with the opponent's intentions. Modern psychology has certainly taught us how one's own beliefs shape the way one interprets reality. Obsession with the other fellow's "evil intentions" and untrustworthiness always prevent assessing his legitimate needs. The existing situation requires the satisfaction of legitimate needs for both sides without viewing them in a warped way. (Rationalizing the opponent as the personification of Evil is a warped way.) The total reduction of fear requires trust and honest relations. National relationships built on trust are not easily attained and are not likely between the U.S. and the Soviet Union in the near future. But demonstrably honest dealings built on open and realistic assessment of legitimate needs can be the basis for reducing fear. As long as deterrence by threat of retaliation exists in its current form and can remain stable, it serves as a basis to solve the longer range problem of reducing the fear and distrust. Disturbance to that equilibrium through more escalation is certain to increase the fear.

Many statements in the press by respected members of the U.S. defense establishment and former government officials indicate a desire to confuse the opponent, create uncertainty, and thus raise his level of apprehension about U.S. capabilities and intentions. This is clearly valid reasoning only if one expects that war is imminent. It is precisely the wrong direction by which to create stability, to pursue peace, and reduce adversarial relationships. We have previously pointed out the one situation in which creating uncertainty has validity. It is in creating uncertainty with regard to the position of retaliatory targets. This uncertainty blunts the acceleration toward first strike capability and thus does assist deterrence without increasing the probability of accidents.

SDI, as is being planned, clearly does not satisfy a legitimate defense need under modern conditions and moreover exacerbates for both sides the underlying real issue which is fear of surprise attack. Arguments which purport to show that SDI increases deterrence are specious and indefensible but, worst of all, they are born of ignorance however well meant. Moreover, deterrence only addresses the fear of deliberate attack, but the probability of deliberate attack is already near zero and needs no enhancing. The proposed SDI does increase the likelihood of accidental and irrational attack and that can only be prevented by stopping the escalation.

4.5.2—BARGAINING CHIPS

An intellectual hatred is the worst,
So let her think opinions are accursed.
Have I not seen the loveliest woman born
Out of the mouth of Plenty`s horn,
Because of her opinionated mind
Barter that horn and every good
By quiet natures understood
For an old bellows full of angry wind?
— William Yeats, *A Prayer for My Daughter*

After SDI was first announced in 1983 and the debate over its efficacy began, one rationale cited when proponents were backed into a corner was that of a "bargaining chip"—a bargaining chip to be bartered in exchange for something equally or more precious.

Two things were erroneously believed by advocates of the system in the beginning: 1) that a leak proof defense was attainable; 2) that to duplicate or just to counter the system the Soviets would be economically or technologically pressed beyond their capability. Neither has proved to be true after even the most cursory analysis. In the beginning the drive was toward understanding the capability of lasers in space as a defensive system. But it is now clear that even though such a weapon is eventually attainable it requires long and tedious development to be compact enough and reliable enough to work unattended in the space environment. So long, in fact, that kinetic energy weapons are being pushed to fill in the gap and time so that adequate countermeasures can be devised by the opponent in the interim.

Such counters would need a relatively simple battle management system plus a powerful radar for acquisition and pointing. In addition, the principles of a laser weapon must first be developed on the ground before a laser can be taken into space in any event. Such development through the experimental stage at one fixed location would be consistent with the ABM Treaty as well. Thus the counter to an orbiting battle station likely can be ready for use before the battle station itself and without violating the Treaty. If such a weapon is stationed on one's own territory and has no orbiting mirror as an "offensive" counterpart, then the installation certainly has a more "defensive" characteristic than an orbiting battle station which can be used offensively.

With this type of reasoning, the SDI bargaining chip takes on the value of a liability. Not only does it not provide the defense promised, but it is also terribly expensive in comparison with a similar ground-based countermeasure regardless of the technology. Further, were the counter a ground-based laser, each can deal with a large number of the more expensive space-based machines. The offending orbiting battle stations further oblige by orbiting one after another in perfect alignment over the "defending" weapon as in a shooting gallery.

With a necessity to counter SDI, the Soviets have been handed a less expensive problem, a less technologically challenging situation, the public relations advantage of being able to say to the world that it is the U.S. that is the aggressor because of orbiting "offensive" weapons. The valuable bargaining chip of SDI from this perspective appears more like a "bad penny" that keeps turning up and no one knows how to get rid of it unless the originators withdraw the idea. The most sensible solution, if one distrusts the opponent, is to give them all the plans for SDI and hope that they try to use it. Let them bankrupt their economy trying to make it work and we take the simple side of the problem. But it might take some lucrative trade agreements thrown into the bargain to induce them to do anything so foolish. Soviet planners are able to analyze the situation as well as we. It is the U.S. defense establishment that seems blind to the facts and consequences in their dedication to carrying out the original marching orders. If the Soviet is afraid of SDI, it is because they realize if deployed, they must counter it by the most reasonable means before it can become operational. That means is to destroy any orbiting elements. And that will bring on retaliation and war.

4.5.3—ECONOMIC SPIN-OFFS

It was through the Second World War that most of us suddenly appreciated for the first time the power of man's concentrated efforts to understand and control the forces of nature. We were appalled by what we saw.

—Vannevar Bush, *Science Is Not Enough*, 1967

If there is any one thing that science, scientists, and technology of the modern era require more than anything else, it is money. To do meaningful research and product development in this period requires exceedingly complicated and expensive equipment. As legend tells it, Sir Isaac did not need a research grant to understand gravitation. The apple was free. Those days have long since departed.

As knowledge has become more precise about the nature of physical reality, humans have required more tools in order to move their point of view deeper into the interior of matter. Even the most basic physical experiments in this era require banks of computers and elaborate equipment that may require years to design and manufacture. Such resources are usually beyond the capability of single individuals. To bring a new idea to light, to test it empirically, to get it accepted and utilized, requires significant team effort.

The notion of the genius working alone in his attic to perfect a brilliant new discovery is almost a thing of the past. There are a few such individuals, of course, but they are often outside the mainstream of science and technology. To go beyond the attic to a wider audience requires time, people, organizations, and above all, money. Thus, teams of scientists, engineers, managers and support staff operate on problems from large corporate bases whereas previously most basic discovery came from the individual and was centered around a university setting. Until after World War II, large corporate and government laboratories doing basic research were almost nonexistent. Today a significant portion of research and discovery, but most particularly applications research, comes from corporate and government endeavors.

Once a team is organized to develop and perfect an idea, they immediately would like job security. After all scientists' kids need to eat and to go to school also. So, after the first project is completed, what then? Organizations must continually justify their existence by new production of whatever they produce in order to provide that security. Even in the ideal state, should one ever exist, individuals and organizations will need to produce at least as much as they consume (plus some for the government) in order to remain in existence for any length of time.

It is wonderfully invigorating for a scientist to investigate whatever seems interesting at the moment, but the dragon of economic reality continues to rear its ugly head. Few are able to go their independent way and pursue only their own intellectual interest. It is the role of management to make certain that the basic production-consumption equation is satisfied even when an organization is just doing research. In fact, in the existing environment, economic justification of pure, non-directed research is quite difficult if not impossible. Only the scientist entrepreneur such as a Lear or Cousteau or Piccard have found the means to remain independent in the modern era. Independence will become increasingly difficult as the cost of ever more complicated equipment is necessary for the probing of life's mysteries and developing the ensuing knowledge into products.

Obtaining the money to remain organizationally strong has forced science and scientists, technology and technologists to increasingly focus on products that meet someone else's need or desire. It is this basic economic issue that is usually the deciding factor when difficult decisions need to be made. The economic realities surrounding modern research drive the decisions as to what should be done. They require the question: Will it sell in some marketplace?

Historically, the chair of the academician-scientist-teacher based in some university was the seat from which new basic knowledge arose. And to a certain extent, that is still true as individual scholars pursue their areas of individual interest and concern. But the universities need modern equipment and machines also in order to stay abreast of new development. Student tuitions cannot cover this need. Thus again, economic necessity enters in the form of directed research and research grants as those who want answers are compelled to pay the universities to find them.

The result in the late twentieth century, as never before in history, is that scientific and technological genius needs to find applications for economic reasons rather than for reasons of fundamental discovery. Thus, basic research has in large measure been replaced with applications research. The alternative is the government-controlled laboratory which is subject to political whim and is equally suspect. Research into saleable applications for technological knowledge and research to solve specific technological problems represent by far the greatest proportion of scientific and technological effort. Products for the military establishments of the world consume a huge amount of this effort. In the mid-1980s, those expenditures exceeded $800 billion per year. It is one of the world's largest consumers of goods and services.[19,20,21,22] (Drug trade is not far behind at $500 billion per year.)

Research into the fundamental nature of things to help better understand our world represents only a small portion of total effort for the period of the 1980s. We must ask the questions: Are the economic decisions that society is making really the rights ones to assure the type of civilization we need? Is channeling research effort into military products enhancing civilization or likely to destroy it?

The nuclear weapons decision taken in World War II continues to have its influence. It was that and subsequent decisions based on a traditional win-lose strategy that is assuring a lose-lose outcome for the world should global conflict again occur. The nuclear weapons genie can never be put back in the bottle. It forces questions about the propriety of future weapons technologies because the introduction of sophisticated military technologies requires the increasing economic stimulation of those industries to maintain the capability even in times of peace. The economics themselves become a driving force.

The technologies now envisioned as a result of further scientific discovery also hold the potential of being equally or more dangerous. Or they can, on the other hand, be used to help stabilize the course of civilization. Nothing envisioned can be as destructive in the near term as unleashed nuclear explosions; but likewise the new knowledge can either enhance the usefulness to humankind of our basic knowledge or be used to further the destructive potential of the power that is already possessed. (Consider genetic manipulation, for example.) The technology itself does not make the choice. Humans guided by their individual knowledge and morality, organizations guided by their collective sense of morality, make those decisions. As should be obvious, there is no absolute sense of morality to guide these decisions. There are only decisions based upon human understanding of the individual or collective best interest. To date, regarding weapons technologies, the decisions have not been very commendable.

The trends of recent years has caused the idea to proliferate that research on government sponsored programs is a gold mine of technological spin-offs for economic benefit. That by using public money for war research, beneficial products can be introduced into the marketplace. This is a valid secondary effect if the initial effort has intrinsic value to begin with. It is not, however, a proper way to create justification for something that has lost or never had intrinsic value. Nor is it a very intelligent thing to do if the basic "game" is producing lose-lose outcomes. Seeding the game with more money just hastens the lose-lose outcome for all, even for those who momentarily believe their win outcomes have increased.

There is ample evidence, now that SDI has had significant amounts of funding for several years, that the economics are starting to drive the system. And when that happens, rationality and morality go out the window. Were there a remote chance that long range benefits could accrue from this activity it might be tolerable as a cure for unemployment. It will keep a small portion of the factories turning. But with the end point being so clearly destructive it is hard to find any justification at all to continue. The notion of U.S. technological dominance of world markets is an appealing one. Ultimate technologies, classified programs, expensive equipment available to favored allies means big dollars. But we can take a few liberties with the words of Marie Antoinette: "Let them eat cake today, tomorrow they die."

4.6—SPACE JUNK

One damning distinction between human and animal culture is that the human leaves a trail of useless debris wherever it wanders.

– Dr. Edgar Mitchell

In all public and congressional debates on the subject of defense systems in space, wars in space, etc., there is one topic that, curiously, is omitted. Curiously, because if one has concern for our children's wellbeing and concern for space as a useful arena for human endeavors, the topic should be near the top of the list of priorities. The problem is "space junk."

In many places in this book, I have suggested that the rules of the game for operating in and utilizing space are different than the rules of the game for Earthbound activities. Without understanding the rules, major errors in judgment are certain to ensue. That is the situation with orbital debris. The problem of space junk is not one of aesthetics. It is not the same as Lady Bird Johnson's "keep America beautiful" campaign of the 1960s. It is a very practical issue about which few political decision makers are even minimally informed.

Objects in Earth orbit remain in their orbits indefinitely unless they are in sufficiently low orbit to be eventually dragged into the atmosphere and incinerated like a meteorite during reentry. Above 150 kilometers from the Earth's surface the decay time for orbits quickly becomes infinite. Thus junk in space is not like a beer can thrown along the roadside. The beer can, or even a bombed-out building, is eventually covered over by the sands of time and reabsorbed into the processes of nature. The Earth can heal its own wounds and recover eventually from this type of human assault, if products are biodegradable. Archaeologists know just how well concealed are the human ruins of only a few hundred years ago. This is not the case for objects in space. Above the very low orbits all particles remain in place indefinitely, large or small. Only the energy from the sun slowly rearranges the orbits of small particles. Dense particles hardly change at all except by chance collision with other particles.

To remain circling eternally in Earth orbit these objects must have energy above orbital energy and less than orbital escape energy. This represents a range of velocities from approximately 25,000 feet per second up to 36,000 feet per second. These are very high velocity objects with significant energies for destruction. Should one be encountered by a satellite or space craft there is a great likelihood of severe damage. Should the space craft and the piece of junk be in intersecting orbits, the closing velocity can range from orbital velocity of 25,000 feet per second up to in excess of 70,000 feet per second. And, as I previously noted, should the space junk have a mass of about one gram or more it is certain that it will penetrate the space craft more explosively than an armor-penetrating shell, spilling its electron blood throughout space. This is the reason that the initial SDI efforts are toward kinetic energy weapons: they are simple and devastating. With modern guidance systems that can seek a target, all one needs to do is hit the target. The kinetic energy of orbital velocity does the rest!

One might protest that "space" is very big and the likelihood of encountering such objects is very small. This is not the case at all. Just a few satellite battle stations destroyed by a Star Wars scenario will provide all the space junk necessary to make Earth orbit a very hazardous place to travel. A few tens of millions of pieces of metal from destroyed satellites, none any larger than a rifle bullet, will make near Earth space almost completely uninhabitable. Unlike the proverbial needle in the haystack you don't need to find them, they find you.

A quick, simple calculation will reveal that there are less than six hundred billion cubic kilometers of space in the sphere above the Earth's surface out to one thousand kilometers into space (620 miles). That seems to be a lot of empty space. However, six hundred thousand metric tons of spacecraft material would provide enough junk to allow a one-gram projectile for each cubic kilometer. That doesn't sound like much in terms of beer cans along the highway. It is an enormous amount if they are whizzing about at orbital energies and one cannot detect them. It is far more than enough to rule out space travel for all except the most fool hardy and to prevent any peaceful uses of low orbits for public use. It so happens that six hundred thousand metric tons is almost precisely equal to 200 of the Apollo-Saturn V vehicles with which we began lunar exploration. Even without being able to exactly know the weight of all 2,500 space craft launched around the world during the first 30 years of space flight, it is estimated that the total amount of mass at launch is about one-half of the amount needed to provide 1 gram for each cubic kilometer of space out to 1000 kilometers. Fortunately, very little of that mass is still in orbit as debris, certainly much less than 0.01%. More fortunately, what does remain is in large pieces, much still in useful operation and not in tiny pieces too small to be tracked. Space engineers of all countries are acutely aware of the problem, going to great lengths to minimize particles remaining in space.

The SDI scenario, on the other hand, proposes a program in which the weight of laser fuel alone could reach 60,000 tons of material, one-tenth of the amount necessary to provide a gram per cubic kilometer. (Laser fuel does not cause a space junk problem because it would be gaseous. It does allow us to roughly estimate the amount of "hard material" because fuel would of necessity be of the order of perhaps one-third of total weight. In the case of kinetic energy devices, rail guns or interceptor missiles, the total amount of matter including solid fuels or energy generators would be added to the amount of potential space junk.) From the section on new technologies we estimated in excess of 100 orbiting battle stations as a minimum. This minimum number alone would supply a significant part of the mass needed to reach the one gram per cubic kilometer mark. By estimating the weights involved by a number of different approaches, one reaches the conclusion that the amount of hard material in space, thus potential space junk, as a result of any unilateral SDI proposal will equal or exceed one gram per cubic kilometer irrespective of what the opponents might do to add to the problem.

Very importantly, exploded material in space doesn't just fly randomly about uniformly filling all space, and then stay in one fixed location like a beer can at the side of the road. All the exploded pieces—if they stay in orbit and don't escape or re-enter to Earth—will periodically return to the same place in the orbit where the explosion took place as surely as salmon return upriver to spawn. They will repeat the journey through the same spot in space every few hours depending upon the period of their individual trajectories. This means that low orbits, under 1000 kilometers, where most of the Star Wars equipment must be in order to be effective in killing ballistic missiles, will have all the particles whizzing by continuously where equipment is destroyed during a space battle.

There are already billions of very tiny particles of space junk in low Earth orbit just from the space activity of the past thirty years. All returning space craft have small pits and nicks in their outer skin just from encountering micrometeorites, paint chips from previous space craft and other tiny pieces of space debris. Each impact of a tiny piece of debris causes from one to several additional pieces as it impacts and gouges out other tiny particles. A bit of debris impacting a larger object acts much like a "cue" ball ramming into a rack of billiard balls on a table. The balls scatter in all directions but if in space would continue to hit other balls occasionally, breaking them into ever smaller pieces.

A simple calculation will show that a space craft with a cross sectional area of only 10 square meters (about twice that of an automobile) could expect to be hit by a particle every three and one-half hours (once every two orbits) if there were one particle per cubic kilometer of space. If those particles were from exploded space stations and uniformly distributed in one-gram pieces across the possible orbital velocities, about one-half of them could impact a space craft at or above orbital velocity, sufficient to penetrate and destroy it. From this one deduces that the average expected lifetime of spacecraft in low orbit after a large battle in space could be of the order of seven hours just because of orbiting debris. What the opponent didn't or couldn't kill, the flying shrapnel would in a short period of time.

Even though I used one gram of debris per cubic kilometer in the example, it is clear from the foregoing that the allowable tolerance for debris is at least four orders of magnitude less than this for continued successful use of near-Earth space. That would mean only one gram of mass for every 10,000 cubic kilometers is permitted in order to assure an expected satellite lifetime of 10 years in orbit. Exploding even one battle station the size of an Apollo Saturn V into one-gram bits would cause space craft survival time in orbit to be about 2 months. In other words, there seems to be no space battle scenario that would not create an unacceptable amount of debris and make low orbits too hazardous to use. If there ever is a space battle, it will likely be the one and only. Replacement battle stations will be destroyed by debris almost as soon as they are put in place. The initial estimates for kinetic energy weapons to be orbited in the early partial deployments of SDI are of the order of 7–8 million pounds, almost precisely the amount required to achieve the foregoing 2 month expected lifetime if they were exploded into small pieces.

This problem has no foreseeable solution. Cleaning up space is not like picking up the refuse along the streets. The small particles cannot be detected, but if they were they would be virtually impossible to clean up. Orbital debris will continue to accumulate regardless of human care in being meticulous. And it stays there forever. The best we can do is be very careful that it accumulates slowly.

Nevertheless, with the best efforts of humans now and into the future, within a couple of hundred years our descendants will face a huge problem with space debris. But if humans deliberately begin to explode weapons, destroy satellites and orbital machinery in space warfare, launching a spacecraft into orbit will be like running through a hail of bullets. To get into a safe orbit up and away from the debris will be like swimming in a piranha-filled river.

The low orbits that are found most useful today, those in the low latitudes above the populated countries, will become the most hazardous. Even children now being born, will essentially be precluded from going into space. Regardless of how severe we think our problems are now, we owe civilization of the future much more responsibility than making the new frontier of space uninhabitable within the first one hundred years that we begin to utilize it.

But even today, the commercial companies who have invested millions in orbiting communications satellites should think twice about cohabitating in space with military weapons that invite target practice. Instead of enhancing security of space, they makes it more hazardous. That limitation is a part of the new "rules of the game."

4.7—GREATNESS OR INFAMY?

We must learn to welcome and not fear the voices of dissent.
We must dare to think about "unthinkable things" because
when things become unthinkable, thinking stops and action
becomes mindless.

– Senator William Fulbright, U.S. Senate, 1964

In section 2.2 we underlined some sentences from President Reagan's speech announcing SDI. In view of the analyses and comments that have raged since that time and presented in this book, it is appropriate to re-examine those underlined phrases with a more critical eye.

It is not difficult to agree with the premise implied in the first phrase: " … dealing with other nations and human beings by threatening their existence." Indeed, mutually assured destruction is not a promising existence. Sooner or later that problem must be resolved. BUT, this is primarily a problem of diplomacy, international dialogue and arms negotiations. Such problems can never be solved by technological means and military measures. The world has lived with the threat for 30 years. If meaningful negotiations are indeed impossible, a few more years of the problem is immensely better than beginning a solution that is guaranteed to make things worse for " … years, probably decades … " before there is any hope of making it get better.

The next underscore is very suspect: "… examine every opportunity for reducing tensions and for introducing greater stability into the strategic calculus on both sides." But this is precisely what has not happened. The words are soothing but totally misleading. The belligerence of the Administration regarding serious and constructive dialogue with Soviets is not well concealed. Hostility bristles underneath the high-sounding words as rocks and shoals beneath a calm sea. It is unworthy of a great nation and precisely what we have complained about regarding Soviet attitudes in prior years.

The Soviet initiative in late February 1987 toward reaching accord in reducing short range nuclear weapons in Europe followed one month later by a breakthrough proposal for on-site inspection of satellite launches were danced around and played down. Only political necessity at home was instrumental in forcing a grudging negotiation. There is utterly no evidence at the operating levels of this Administration of a wish to achieve long range stability through negotiations.

The next underscore is: "If the Soviet Union will join us in our effort to achieve major arms reduction ... ." If this statement were sincere and could be realized, then the SDI is not required. Proceeding with SDI and wanting arms reduction are inherently prima facie contradictory ideas.

The next statement is a major point: " ... with measures that are defensive." I think I have amply made the case that there are no systems in modern warfare that are just defensive! It is a traditional but simplistic and misleading term, inappropriate for modern technology.

Consider the next crucial phrase: " ... destroy strategic ballistic missiles before they reach our own soil or that of our allies?" This sounds wonderful in a ringing, soul stirring speech but not when you look below the surface! The professionals have known for ten years that the numbers game was out of hand and could not be technologically contained. That the technology to be used would attempt to destroy missiles in the sky above Soviet launch facilities. We may think that is wonderful. They are not apt to think so! As one presidential advisor remarked, SDI is not planned as "a shield over America but a lid over the Soviet Union." Now, reverse the situation. Are we prepared to let the Soviet Union, or any nation, unilaterally deploy a technology into space that can destroy not only missiles but ANY vehicle in the skies above our nation? I think not! We can only prevent the Soviet Union from deploying their own similar technology, should they choose, in the same way that they can prevent ours—shoot them down as they pass overhead! It is the same as a mutual naval blockade of each other's harbors.

Let's try another phrase: " … worth every investment necessary to free the world from the threat of nuclear war?" When the Pandora's Box of nuclear weaponry was opened in World War II, the dream of removing that threat was forever rendered hopeless. Every physics student at the university level knows the rudiments of making a nuclear weapon. All that is required is the skill, time and some stolen fissionable materials. Technology is incapable of undoing what it has already done. The solution to eliminating existing nuclear weapons is no longer technological. To do so requires cooperation among the nuclear powers to police the world against terrorists and rogue nations who would use nuclear blackmail against civilization. SDI presents the opportunity to learn from the nuclear lesson: don't deploy a technology that will be regretted 25 years hence!

A most blatant misrepresentation is the next one: " … consistent with our obligations under the ABM Treaty … ." There is nothing in the SDI approach that has been consistent with the ABM Treaty of 1972, nor can it be. One point defense installation is permitted by the Treaty along with a limited amount of ground-based research and testing of new systems. Prima facie, SDI as announced will be an overt violation of that treaty as well as every intention behind the SALT agreements. The intent to subvert or to "reinterpret" the treaty has been demonstrated by the Administration at every occasion since 1983. But the rationale has been paper thin. The Administration's claim regarding Soviet "bad faith" and willingness to violate the Treaty, thus making it necessary, was disputed in Congressional testimony by two of the negotiators for the Treaty: Lt. General Royal Allison and Ambassador Garthoff, both of the U.S. ABM Treaty Negotiating Delegation. They testified on March 11, 1987, before a joint committee meeting of the Senate Foreign Relations Committee and the Senate Judiciary Committee that the treaty was representative of both sides' intent and that there was no indication of Soviet violation of the Treaty except on the issue of building radar. Both sides, Garthoff testified, had built radars that if used for ABM purposes could be construed as violations. If used for other purposes such radars would not necessarily be violations. Each side has accused the other of violation in this regard. Thus there seems to be some ambiguity in the provisions for developing radar, loopholes which both sides have interpreted in their own way. The remainder of the Treaty is very specific and precludes SDI activities beyond laboratory type testing of new ABM weapons (see Appendix A2). The President's statement itself, of intending to remain within bounds of the Treaty is manifestly contradictory and not the policy that has been pursued. The only way the objectives of SDI can be legally met is to renegotiate or to withdraw from the ABM Treaty.

And then another: "We seek neither military superiority nor political advantage." It would seem that the opponents might have difficulty believing that since a relative parity currently exists. They would probably adhere to the adage: "What you do speaks so loudly that I can't hear what you say." Whether or not the opponent is trustworthy himself is not the point. He has looked at the words and the actions from the point of view presented in the foregoing. They can only increase his alarm and thus decrease OUR security as well as his own. Trustworthy or not, the adversary is not a fool. We should not be surprised at his concern, nor at any activity he undertakes to increase his perceived need for security. Our alarm would be equally as great or greater if the initiative came from the other side. But actually it does not matter from which side an escalation commences, it is the fact of escalation itself that creates the danger, first to one then the other, but always to the entire world.

Finally: " … an effort which holds the promise of changing the course of human history." And indeed it will. There is nothing more certain than this promise. But the direction of change is most likely to shorten the prospects for a future history, not improve them as one might hope. Lose-lose continues to gain on the other outcomes.

There are four major sectors of society so far affected by this sweeping announcement of revised national priorities. They are the military forces, the research community, the aerospace armaments industrial community, and areas of the executive and legislative branches responsible for oversight and funding. These are the only areas of society which stand to benefit in the short run. Civilization stands to lose in the long run. As pointed out, however, large numbers of scientists see the grimness of the overall reality and refuse to participate. Some, however, want to see the research continue on weapons technologies, but not deployment and utilization of the system. This is likely short sighted, also. If the technology exists, it will be used, now or eventually, unless agreements can be made to the contrary suppressing that path of development.

While those doing the paddling (swan analogy again) seem to have paddled off in their own directions and while those sectors most affected by the increased budgets moved to justify additional funding, the President has clarified some of his ideas during the years following the original announcement. Some more recent quotes should be noted.

President Reagan:

"We won't put this weapon—or this system in place, this defensive system, until we do away with our nuclear missiles, our offensive missiles." "But I do know what we're going to do, and I have stated it already, we would not deploy … It is not my purpose for deployment—until we sit down with other nations of the world, and those that have nuclear arsenals, and see if we cannot come to an agreement on nuclear weapons."

– *Defense Daily*, 5 November 1985, p. 17

"If and when we finally achieve our goal, and that is a weapon that is effective against incoming missiles … then rather than add to the distrust in the world and appear to be seeking the potential for a first strike by rushing to implement, my concept has always been that we sit down with the nuclear powers, our allies and our adversaries and see if we cannot use that weapon to bring about … the elimination of nuclear weapons." (If that conference failed to reach an agreement for mutual use of the defensive system,) "we will go ahead with deployment."

– *Defense Daily*, 7 November 1985, p. 40

However, a scant four months later, the President is seeming to say something totally different:

"But innovation is not enough. We have to follow through. Blueprints alone don't deter aggression. We have to translate our lead in the lab to a lead in the field. But when our budget is cut, we can't do either."

– *Defense Daily*, 28 February 1986, p. 326

And later in the same year:

"Well that isn't exactly what we've proposed to the Soviet Union, delaying our Strategic Defense Initiative."

– *The Washington Post*, 13 August 1986, p. A14,
Transcript of President Reagan's news conference.

But the organization responsible for carrying out the directive has never seemed to be confused about its goal:

Lt. Gen. James Abrahamson, Director SDIO:

"I have consistently said that we're driving for an early 1990s decision."

— Defense Daily, 1 July 1986 pp. 5–6

" … if our great progress continues, we could have a decision on deploying SDI earlier than the presently expected 1990s.

— Defense Daily, 24 March 1986, p. 122

It might appear from the foregoing that the swan has been carried to wherever the "furious paddling down below" is taking it. Very clearly what has been said and done in 1986–1987 is quite different than the earlier pictures that were presented. Two points are dramatically clear to all who have looked seriously at the technology for four years: 1) There is no hope of a population defense, and 2) the project is highly destabilizing. Yet until that is openly acknowledged at the highest levels, the people will continue to be confused and deceived by the memory of those early promises.

Even more dangerous than the contemporary (and hopefully temporary) confusion of the public and the Congress is the concerted effort to make the SDI irreversible in spite of its technical and strategic flaws before the end of this administration and before the fog of confusion can be lifted. Indeed confusion and controversy permit the funding to continue, so that the point of irreversibility is ever more closely approached. There is no lack of evidence that such is the orchestrated plan. As early as 1983, a strategy was proposed and circulated through the Administration with such a thought in mind. The strategy was based upon a paper commissioned and circulated in the Administration by the High Frontier foundation. It was entitled: "A Proposed Plan for Project on BMD and Arms Control," by John Bosma, editor of *Military Space*. This paper was brought to the attention of the Senate Foreign Relations Committee in May 1983 and was discussed in *Harper's Magazine* in June 1985. The objective listed in the paper was:

"… To get an early IOC BMD program underway as soon as

possible and to develop enough political support for this BMD effort to ensure a) its initial startup during the second term of the Reagan presidency, should that transpire; (b) sufficient political visibility and/or organizational and programmatic momentum that it could not be turned off by a replacement or successor Democratic Administration." ((IOC is the acronym for Initial Operations Capability; BMD)

That such a strategy has been and is being followed by those "doing the paddling down below" is hardly in question. This was reaffirmed in January 1987 by the Secretary of Defense's announced plan to move toward an early partial deployment of SDI and by Attorney General Meese's comment as reported in the same month in *The Washington Post*:

"Yet another motive, for which no decent argument can be made, is that it would lock in future presidents as well as implanting a firm Reagan legacy for history. What else could Attorney General Edwin Meese have had in mind when he told the Washington Yale Club recently that the advantage of a quick first phase deployment is that ... (Meese)... "it will be in place and cannot be tampered with by future administrations"?

– Philip Geyelin, *The Washington Post*, Jan. 8, 1987

Were the four years of study and research to have shown that the Strategic Defense Initiative had merit and had a reasonable chance of enhancing security, such political maneuvering might be tolerated. There is nothing like being proved to be right in one's assessment to reinforce power and to obtain favorable historical comment. But having made an impulsive, dramatic maneuver which is totally and completely in error, then continuing to press forward in spite of the evidence of the error can only assure a legacy tainted with infamy.

CHAPTER 5—NEW RULES FOR CARE AND FEEDING OF FOXES

5.1—SUMMARY AND CONCLUSIONS

Of making books there is no end, and much study is a weariness of the flesh. Let us hear the conclusions of the whole matter ...

— Ecclesiastes 12:12–13

I will summarize in bullet statements the conclusions I have drawn from an analysis of the SDI effort. They are all negative.

I searched diligently to have something good to say if for no other reason than to avoid accusations of being irrationally and emotionally biased but found no reason except that it keeps the factories running. That reason seems totally inadequate in the face of the dangers. There are alternatives for gainful employment. The principle means by which one can view and accept the arguments and conclusions of this book is to abandon the premise that humankind is forever destined to incessant warfare. Since civilization evolves under human guidance, let's evolve in a sane direction.

I would prefer to speak admiringly of our President, our government and our military in their decisions and their efforts in the public interest. In this case I cannot do so. I must oppose with all my strength an effort that I believe is damaging not only to our country but in the larger sense to the entire world and certainly to future civilization should it exist.

I began opposition believing that well intended but misguided advice was driving the move toward putting weapons in space. As the momentum increased for the program, the professional objections have also increased but the population remained uninformed, largely following President Reagan's direction without too much question. The polarization of opinion has deepened. Those with a financial interest in SDI, those with belligerent inclinations toward the Soviet Union and those with traditional military instincts are moving strongly within the current political environment.

All others who have studied the problem are motivated to resist. The public must become aware and interested in the outcome because of the dramatic irreversible impacts on the future. The arguments of supporters for SDI have changed continuously as first one and then the next of the original justifications have been shown to be invalid, but the momentum has hardly slackened. There is clearly an element and an influential one at that, whose attitude is "don't confuse me with facts." It would seem that lack of attention, direction and control of the military establishment by the Commander in Chief has parallels, but in more dangerous fashion, to the runaway activities of the National Security Council during the late 1980s. When foxes guard the henhouse, someone must watch the foxes, else only they are satisfied.

There are no valid arguments to counter what has been presented herein. The only valid objection is: "What are the alternatives?" Those we shall present in the following sections because there are viable alternatives for the future other than the ones into which SDI takes us. We happily present them realizing one cannot justly criticize without presenting a realistic better option.

THE FACTS

◊   No scenario of SDI weapons mix can reliably produce more than 95% protection against nuclear missiles currently—and likely never.

◊   The leakage of 500 (5%) of existing opposing warheads, or even 5 warheads is a totally unacceptable number from any point of view except the military whose policy is deterrence through retaliatory capability.

◊   Because of submarines, retaliation capability is assured after the first strike without SDI. The numbers may change some but not enough to drastically alter the overall retaliatory outcome. The population is lost with or without SDI.

◊   First strike capability (accurate guidance systems plus sufficient numbers of missiles to attack all sites), can be countered by retaliating before warheads land, thus retaliation is not lost even with first strike targeting. The cost is a decrease in decision time. The population is lost in either case.

◊   Whatever effectiveness number is used for a defensive system in space, it can be countered by saturation at equal or less cost.

◊   Weapons in space reduce warning and decision time toward zero.

◊   Laser weapons reduce warning time to zero providing a weapon of ultimate surprise.

◊   Laser weapons are not defensive, they can equally easily be offensive weapons of mass destruction.

◊   The orbiting of laser systems causes preemptive first strike to be the only defensive option because lasers are a zero-warning time weapon.

◊   No aggressor will deliberately use nuclear weapons massively as a weapon of conquest because they render the public assets useless. An aggressor might use space-borne lasers because they only selectively destroy public assets.

◊   The introduction of any weapons into space will create an escalation toward lasers and/or nuclear explosives in space.

◊   Nuclear missiles must be held by both sides if lasers are believed to be potential weapons in order to prevent, by threat of massive nuclear retaliation, surprise laser attack on the population or on the space defenses. Nuclear missiles are more easily justified with lasers than without.

◊   Ground-based lasers can be made operational within the next decade which can destroy any orbiting satellite or defensive system at a fraction of the SDI costs.

◊   A "defensive" preemptive strike against orbiting weapons will almost be mandatory during any period of major crisis between superpowers.

◊   Parity of weapons between opponents is the only stable condition to deter weapons of mass destruction.

◊ The side that first begins to improve its technology and thereby threatens the other side's legitimate defense needs places its population at greatest risk from a preemptive strike.

◊ In this era of computerized warfare with weapons of mass destruction, the risk to the population and its assets is greater than the risk to the military assets. The military operating the controls of war are protected from direct exposure, the population is not. Defense policies must be based on population protection not military needs. The two are different.

◊ Deterrence is not a policy which can prevent accidental or irrational war.

◊ The weapons of both powers are approximately equivalent and of intercontinental range, therefore the legitimate military needs for defense of each are the same. The difference between aggressor and defender is only in the intent. Intent has no technological solution.

◊ With nuclear missiles and lasers as weapons, and with space as mobility medium, the technology leaves no margin for machine nor human error.

◊ Moral or social justifications are moot in the foregoing condition, the technology becomes the danger and humans lose control.

◊ Secrecy, confusion and deception of the opponent are only valid if war is imminent. It is not valid for peacekeeping. Such tactics display aggressive intent, cause apprehension to the opponent, and increase the likelihood of preemptive first strike. If near zero warning time weapons exist, these tactics will almost certainly assure an erroneous or accidental preemptive strike.

◊ Orbiting kinetic energy weapons reduce decision time to 30 seconds and threaten the ability of any nation over which they orbit to conduct legitimate high altitude and space activities. They are no less onerous than lasers in this regard.

◊ The "rules of the game" for space military activities with near zero warning time weapons are vastly different than pre-space warfare. Surveillance and communication alone are stabilizing space activities. Everything else is destabilizing.

◊ Any space war is most certain to render near Earth space useless for peaceful activity by making it too hazardous for humans and too high risk for commercial satellites.

◊   Traditional military strategies seeking win-lose outcomes produce lose-lose outcomes for the populations with the use of nuclear and space technologies.

◊   The correct statement of the contemporary military problem is not the "need for defense" but rather "freedom from fear of surprise attack." Both of the words "fear" and "surprise" offer options for dealing with the issues rather than by escalation of weaponry.

◊   Nuclear explosives have no defense; space in not defendable; lasers provide ultimate surprise. Nations have no alternative but to learn to settle differences by other means.

5.2—A THOUGHT

What is the first business of philosophy? To get rid of self-conceit. For it is impossible to begin to learn that which he thinks he already knows.

— Epictetus, *Discourses*, Book. II

The author has severely criticized this program named the Strategic Defense Initiative and the individuals who perpetuate it. It is a very bad program for all the reasons cited. That does not mean it should not have been considered. Indeed the error is not in attempting a bold and creative initiative to change the balance of terror in the world. That was most commendable. The error is in continuation of the program after it has become obvious that it cannot begin to meet the original objectives, indeed it takes us in the wrong direction. It is a flaw in the decision-making processes of government that permits the error to perpetuate itself.

Having made the exercise of evaluating the specific technologies available for military purposes and reviewing the history of warfare on our planet, one must be profoundly impressed at the way Nature has arranged itself. The particular laws of physics applicable to this situation: the energy within the atom which powers nuclear explosions, the speed of light upon which laser effectiveness depends, the gravitational constant of the universe which makes getting into space the problem that it is, and the vastness of space itself, all have combined in such a way so as to make it impossible for humans to export their warlike tendencies into space without committing suicide for the species. Rational beings will not do that and will therefore change their behavior. If we are irrational, then the Universe doesn't need us, and we self-extinguish. It is a very impressive and profound realization. We will get just what we deserve through our own choices. That seems to be the evolutionary lesson.

And with that thought in mind, perhaps we can be grateful for being forced to look at the choices directly. It should make us see that the future cannot be like the past in this regard.

5.3—POT POURRI

The man who is tenacious of purpose in a rightful cause is not shaken from his firm resolve by the frenzy of his fellow citizens clamoring for what is wrong, or by the tyrant's threatening countenance.

— Horace, *Satires*, book III, 23 B.C.

There are several topics that, while not directly tied to the SDI program, are peripherally related and give cause for concern. These have to do with the general emphasis on armaments and military buildup. Whereas the basis for a people's security and well-being historically resided in the strength of their defensive preparations, we are observing in the modern arena of strategic confrontation that the existence of such weaponry even for defense poses unacceptable threats to life for those whom it purports to defend. In terms of historic trends this lesson should not be neglected. It seems to be telling us that the traditional basis for security, military preparedness, is not the unalterable wave of the future. That is not to mean that military readiness can go away or that it has no future at all. Indeed as long as the veneer of human civilization is so very thin and we resort to armed confrontation with small provocation, a strong military presence is necessary for a strong nation. But we must be searching for better ways to describe relationships between peoples and nations and to resolve conflict within those relationships short of armed confrontation. We have learned to do so within civilized nations, why not between nations as well?

The concepts of self-determination, human rights and democratic principles have propagated significantly in the twentieth century. Never before in history have democratic principles been so embraced or even given lip service as in the modern era. But democratic principles and war making cannot survive together. Waging war, even in the name of defense, cannot be done democratically. War is an activity requiring authoritarian control and hierarchical command structure. As soon as a democracy votes to go to war, democratic principles must be restricted. They are too inefficient for war making. The fledgling attempts at democracy throughout history all perished when the democracy was not re-established following a period of war. Should nuclear war erupt from existing or future confrontations it is most unlikely that democratic forms of government could be re-established among the survivors. The laws of the jungle would most certainly prevail for many generations.

The people of the North American continent have not been subjected to warfare on our homelands for over a century. We therefore tend to forget how fragile and tenuous are the threads that hold together the principles of self-determination and human rights. To maintain those principles there can never be doubt that military activity is firmly under civilian political control and devoted to the best interests of the population. It is precisely the strong police and military influence on Eastern bloc societies that causes concern in the West. If military dominance of space activities is permitted to occur, and most particularly if weapons are permitted into space, democratic institutions are in severe jeopardy. Should the knowledge of how to utilize the space environment reside primarily in those programs devoted to warfare, managed by those institutions and those individuals charged with defense, then military dominance of space will be complete and through it the possibility will exist for world domination by some tyrant.

When President Eisenhower enunciated the open skies policy and established a civilian space agency to explore the peaceful uses of space, he foresaw the problem of the dominance of space activities by the military-industrial complex. We are now confronted with that possibility, and it is not one that should be permitted to continue. During the decade of the 1980s, large numbers of high-level positions in NASA have been occupied by career military officers on active duty. These are all devoted and competent gentlemen, but they are first and foremost Military officers. They should not command a civilian agency. The astronaut program which, in the early days of space flight, drew heavily on military test pilots as astronauts has turned increasingly to civilian sources for personnel to broaden the base of skills. That program had, until recently, epitomized the peaceful exploration of space and consciously de-emphasized the military background of astronauts. That is no longer the case. The Manned Spacecraft Center at Houston has become in practice another military base, as many who have been there have noted. The Reagan Administration's de-emphasis of the commercial role and the scientific role of NASA in order to support its military goals has made, in effect, the NASA budget a hidden part of the budget of the Department of Defense.

Space projects have a difficult time finding support unless they contribute to military goals. The military services have sought from the earliest days of space flight to dominate that arena. They now seem to do so. The DOD budget for space activities is now twice that of NASA and the NASA budget must be viewed as mostly in support of military needs. This seems to be a most unhealthy trend for those devoted to personal freedom, human rights, and democratic processes.

5.4—BETTER SOLUTIONS

To criticize without pointing to reasonable alternatives is little better than dogs barking at the night.

                                              – Dr. Edgar Mitchell

There would seem to be only one approach that could be worse than SDI and that would be to start the war immediately and get it over with, sparing everyone the suspense of waiting for the "other shoe to drop." But if we want positive solutions to the problem, there seem to be several. They need to be discovered within the new rules of the game for nuclear and space warfare. Positive solutions means those solutions that reduce the risk of war, not increase it. The need is to reduce the probability of lose-lose outcomes.

    The first step in finding positive solutions to the problem of national security is to redefine the problem statement from that of "defending the nation" to a statement of "achieving freedom from the fear of attack." Specifically in the current moment with SDI, the problem is fear of surprise or accidental attack as a result of weapons escalation in space. In the longer-range view it would be "fear of any attack" that needs to be addressed. This problem statement resolves the difficulty whereby a policy of deterrence allows the likelihood of accidental or irrational attack to increase. It resolves the conflict between military and population interests. Finding a way to protect the population was clearly the President's original intent, but the right solution did not evolve because of traditional win-lose implementation of the activity.

The new statement is a better military policy in general than is deterrence. All wartime activity is designed to contain the fear on one's own side while endeavoring to increase the fear on the other. Increasing the opponent's fear is an appropriate activity if one is at war but is inappropriate if one is seeking peace. Thus a very clear distinction in military activity between peacetime and wartime activities under the new problem statement is whether activity is designed to minimize or maximize the opponent's fear. This gives a very sensitive test to military management policy. It also provides a "litmus" test between defenders and aggressors. Because technology has brought us to the point of being unable to distinguish between defensive and aggressive military activity in traditional ways, a new test is necessary. Observing the "fear producing" activities of one side or the other provides such a test. It is a litmus test of "intent,"which is the only criterion by which one can distinguish between aggressors and defenders.

(It is interesting to note that in primitive societies a tactic often employed before the battle was to dress the warriors in animal skins and horrifying masks representing the evil spirits, then to make loud ghoulish and shrieking noises while brandishing weapons in order to strike fear in the heart of the opponent.)

There is only one difficulty with a "fear" problem statement. It is viewed as negative, depressing. Further, it is not "macho," and to be warriors we must, of course, be macho. People resist talking about fear just as they resist talking about death and funerals. Such a reaction is irrational, and needs to be overcome, since the "freedom from fear" problem statement provides the most sensitive, accurate and practical measurement device to assure moving toward a peaceful solution to conflict. We will address the question of measuring fear in the following paragraph. (There are also those types who will say; "We ain't afeared o' nothin." They must be ignored because they are lying.) The new problem statement forces recognition that the issue is bi-lateral, indeed multi-lateral in a peacetime environment. Multi-level solutions cannot be achieved in the current instance without first resolving the issues between the U.S. and the U.S.S.R., however. Both opponents must achieve simultaneous solutions to the problem or neither can, even though there can be unilateral activity toward fear reduction. The fears of one side, if they result in military buildup, translate into fears on the other side creating escalation which only exacerbates the problem. The traditional military posture of maximizing the opponent's fear (or being unconcerned about it) has always assured that escalation would increase, and wars be almost continuous throughout history. If one is seeking peace, even unilaterally, the only policy change required is to minimize the opponent's fear while maintaining an adequate arsenal to counter his attack. A maximum arsenal is not required, only one sufficient to avoid attack. In the current instant the lowest MAD level to unacceptably damage the opposing population would be sufficient. One nuclear warhead per major city is enough. (This policy should translate into a modern equivalent of Theodore Roosevelt's "Walk softly and carry a big stick." Perhaps it could be: "Cooperate but keep a small nuclear arsenal.")

With the new problem statement one can easily discover two of the four independent reasons why SDI is doomed to failure. Even though it was a noble proposal to reduce the fear on one side, had it been successful it would maximize fear on the other because it proposed to defeat legitimate defensive needs and threaten routine operations of the opponent. The other reason, of course, as has been demonstrated, is that the technology is self-defeating with regard to surprise attack, which elevates the rational fear level on both sides. It increases the risk of surprise attack rather than diminishing it. The third and fourth reasons not discernible from the new problem statement are that SDI is more economically countered by the opponent than duplicated and that it fails to overcome the "numbers game."

5.4.1—MANAGING AND MEASURING FEAR

When fear seizes, the mind's rational thought departs. When it lays unnoticed beneath thought it warps judgment. Only when fully acknowledged and confronted can it lead to rational action.

– Dr. Edgar Mitchell

We can organize the concept of fear management into two categories: those of rational and irrational fears. Rational fears are those grounded in testable and verifiable events. They have a basis when the measurable probability of an undesirable event is unacceptably high or verifiably begins to increase. Irrational fears, on the other hand, are not quantifiable or verifiable. They come from two sources: 1) unverifiable speculation about undesirable events, and 2) unknowledgeable or biased individuals irresponsibly sowing the seeds of fear. All humans, even very rational ones, are subject, from time to time, to moments of irrational fear, but that is not the issue. (There are also, of course, totally irrational people but we will ignore that category in the current instance.)

Fear of individuals who are unduly subject to any category of irrationality becomes a rational fear for a constituency when such individuals attain positions of authority. In the current instance regarding SDI, in addition to rational individuals playing by archaic strategies and thus arriving at no win outcomes, there seems to be a significant amount of irrationality being spread around as well.

Rational fears must be managed through appropriate policy decisions and positive action. Irrational fears must be dealt with as personal problems of individuals. They have no place in the administration of public business. Unfortunately, biased and unknowledgeable persons seldom recognize their own irrationalities and become an acute problem when in influential office. No person is able to accurately discern by themselves all of their irrationalities and thus the human condition exists with a certain residual amount of irrationality as background "noise." Only knowledge and willingness to examine personal belief can help reduce residual irrationality to a minimum. A true danger arises when a few biased and misguided individuals with irrational fears attain positions of power and impose their thinking on a group. The result is at best confusion and at worst mass hysteria. Many of the pro-SDI arguments seem to provide fuel to stimulate this irrational process.

5.4.2—APPLYING THE NEW POLICY

None so blind as those that will not see.

— Matthew Henry, Commentaries, 1708–1710

Let's use the foregoing notions to examine the existing situation of conflict between the superpowers:

Because no rational aggressor will use nuclear missiles in an aggressive mode in face of the unacceptable damage to his own or the world's assets, the fear of nuclear surprise must stem from accidental launch or irrational decision-making regarding defense. Threat of even minimal nuclear retaliation would additionally assure that an aggressor's nuclear attack will not occur. Thus, fear of a deliberate nuclear attack by either side is an irrational fear. Even though a MAD condition is undesirable, it is in existence, and it is effective in preventing nuclear aggression with rational leadership in power. (A caveat, which must be added to the foregoing statements, concerns the extent to which some individuals within either or both groups of champions still harbor the false notion that limited military objectives can be obtained with nuclear strikes. As long as such plans exist, we cannot say that the probability of deliberate attack is absolutely zero. I will, however, relegate such thinking to the category of "irrational decision-making regarding defense." It is the kindest thing I can say about such plans.) The problem to be overcome, then, from the existing alignment is fear of accidental or irrational attack. This fear is a rational fear for both sides. And the basis for fear can be computed quite accurately in the form of a probability of the undesired event occurring. This fear is currently linked to situations of international crisis, which in turn stimulates the alert levels of the respective militaries.

One solution that could be implemented is to keep the alert level of both sides at a minimum so that fear levels of nuclear surprise remain at a minimum. Uncoupling the nuclear force alert mechanisms from other issues of political and diplomatic importance would be a move in the correct direction of reducing fear levels. This de-linking would effectively remove strategic nuclear weapons policy from the other arenas of international confrontation between the superpowers. Weapons policy could then be managed as a separate area of discussion. De-linking could be done unilaterally without adverse effect to retaliatory capability! This is true because deliberate use of strategic nuclear capability to enforce other areas of national policy cannot be taken seriously, not if one's own population has any value—it is not rational and cannot be perceived rationally by anyone. The linkage only causes alert levels to increase as a side effect of crisis, which in turn causes the danger of accidental or irrational surprise attack to increase likewise. In other words, saber rattling with nuclear weapons is not credible. Thus there is no rational fear from this event. But the probabilities of an accidental or irrational attack do increase as long as alert levels are linked to the international scene. This is a rational fear. Alert levels of retaliatory forces should only be linked to the alert levels of the other side, not to anything else, in order to keep rational fear at a low level.

Clearly either conventional military or nuclear force build up also causes the fear level on both sides to reasonably increase. But as long as strategic balance is maintained and superiority is not a goal of either side, fear levels are manageable and in balance. SDI caused fear to increase significantly on the Soviet side before thorough analyses could be done to show its technical flaws. Analysis has shown that the Soviets should have little to fear from SDI, but rather it is the U.S. that should be fearful of what it is doing. The U.S. has the option of stopping its development, the Soviets will only have the option of destroying SDI battle stations if they are deployed or of submitting to the threat to its sovereignty. But both should be equally fearful that weapons of ultimate surprise in space, or even those that reduce decision time, will trigger the war that neither side wants.

With the new statement of the problem, it is axiomatic that fear on either side feeds fear on the other. Thus parity must be maintained to push fear levels toward a minimum. Military hegemony is not a valid goal within the new problem statement. There is, however, a certain tolerance range within which parity can be maintained. Neither side can give up its capability to massively retaliate without its own fear level increasing. It cannot get too far ahead without the opponent's fear level increasing. As long as the probability of deliberate attack is kept near zero by the MAD mechanism, rather wide variation can be accommodated in the actual levels of weapons, both nuclear and conventional. (It has remained near zero for about 30 years in spite of changes in missiles and warheads.)

The fear of accidental and irrational attack then dominate the fear levels and the probability of war as long as MAD exists. The rational fear of accidental attack is directly related to decreases in warning and decision times, to the increased automation of systems, to the decrease in response time of systems, to the alert levels of the system, and is inversely related to the reliability of systems. The probability of accidental attack can be calculated rather well from failure mode formulae and response time data. The probability of accidental attack should be very near zero to keep fear levels low. Highly automated zero warning-time weapons in space, however, bring that probability almost to 1.0, many orders of magnitude higher than it should be or need be.

The probability of irrational people controlling one or the other system is less easily assigned. It is however in the best interest of both sides to make certain that their respective command and control systems are free from unstable individuals and that the system design minimizes the risk of irrational activation. The system susceptibility to irrational acts can be calculated as a probability. The design should also make this probability very near zero.

With the new problem statement it becomes clear that secrecy and covert maneuver also increase fear and decrease stability. Likewise, activity that is contrary to announced policy increases fear. Words must match deeds in order to reduce fears and the words need be reassuring. These statements present large potential for changes in policy decisions and actions toward those that would be stabilizing and thus fear reducing. The activities of both superpowers have failed dismally in the foregoing ways. It is abundantly clear that the negotiating tactics of the superpowers suffer from lack of "good faith" negotiations under the test of reducing fears. Both sides are equally guilty because they: 1) play by win-lose rules, and 2) are subject to much political and public posturing. But keeping one's words and one's activities speaking the same language is a unilateral decision that needs no consensus and that eventually results in successful negotiations if coupled with understanding of both sides' legitimate needs. Not to do so serves no useful purpose and increases the fear levels. The new rules forced by the new problem statement not only work in principle but work in practice because they consistently produce win-win solutions in problems of human dynamics. Although increasingly used in private dealings successfully, such strategies seem never to have found their way to national levels.

In the foregoing regard, it is obvious that on-site inspection and verification of weapons developments are important steps in reducing fear. They are not, however, critically important to the actual strategic postures. Each side knows within very small margins of error what is the other side's actual and planned capability. Claiming that they don't or that the other side is ahead is within the old rules of the game for propaganda and for gaining favorable funding decisions but is counterproductive for reducing fear. Satellites and covert intelligence operations are quite effective in keeping each side informed. There is little of strategic importance that can be kept secret for very long even in the Soviet Union and certainly not in the United States. Thus, much of the cry for verification is window dressing and game playing within the accepted "norms" for distributing propaganda.

Inspection and verification, however, are important good faith moves that can demonstrate peaceful intentions and reduce significantly the capability to manipulate irrational fear even though it would only increase the accuracy of knowledge by at most a few percent. This can be done unilaterally without damage. Open inspection and verification would also dramatically reduce dependence upon covert intelligence and would thereby help eliminate other areas of superpower confrontation: the frequent need to banish members of the other's diplomatic corps and to "bug" each other's embassies. Such events are traditional exchanges and actually make little sense. They are just a part of the diplomatic game between adversaries.

The rationale should be, if peaceful intent is the policy: They have it or can get most of the information anyway, therefore openly make information available and reap the benefit of an improved negotiating climate and world acclaim. In this regard, since demonstration of peaceful intent is the key to stability in the modern world of superpower military confrontation whether one uses new or old rules, it is important that intent be clearly demonstrated by both sides, if indeed peaceful intent is the intent of those in respective positions of power.

5.4.3—CONVENTIONAL FORCE CONFRONTATION

The nature of the universe is such that ends can never justify the means. On the contrary, the means always determine the end.

— Aldous Huxley

Let's turn our attention away from strategic weapons for a moment and apply the foregoing ideas to other areas of conflict. In these other areas, conventional military power plays a supporting role in national policy rather than the central role which nuclear power currently commands in superpower relations. It is likely we will need to amend or add to the problem statement in order to reflect this different and more traditional role. Achieving "freedom from fear of attack" is a necessary but not sufficient statement to reflect the entire scope of the military in modern superpower relations. There are also roles associated with aiding allies, assisting in times of disaster, being "ambassadors" to foreign nations and preserving the peace, to name a few.

It is clear that the superpowers dare not risk direct military confrontation even with conventional forces anywhere in the world for fear that conventional engagements would escalate into nuclear engagements. If defeat were imminent for either side, a rational fear of defeat could produce an irrational use of nuclear weapons. Thus a policy of keeping fear levels at a minimum dictates the need to continue to avoid direct military confrontations between superpowers or between any of the major allies of the superpowers, specifically between NATO and Warsaw Pact nations. Current policy in this regard is consistent with the new problem statement. As previously stated, safety could be enhanced even more by an announced, even unilateral policy, of de-linking nuclear force alerts from conventional military activities. But there are other issues as well which are mostly embodied in the discussion to follow.

5.4.4—IDEOLOGICAL CONFLICT

The imperatives of technology and organization, not the images
of ideology, are what determine the shape of economic society.

– John Kenneth Galbraith, *The New Industrial State*, 1967

To go beyond John Galbraith, it is clear that underlying ideological beliefs do play a role in creating risk of armed conflict, because of emotional political and ideological rhetoric used to persuade the populace plus organization structure founded upon some underlying ideology. This is not of necessity true but is a fact of the contemporary alignments. Strongly held ideological differences in confrontation will of necessity create conflict, therefore the question is not one of avoiding conflict but one of managing the conflict short of using military power. Only the human tradition throughout history of appealing to arms as the final arbiter of national conflict provides the reason for armed confrontation. But because the foregoing is true and because serious international crisis can escalate to promote armed confrontations and nuclear alerts, we need to find new rules of the game for this aspect of the problem.

The west is dedicated to the ideas of the rights of the individual and to the principle of individual economic betterment through capitalist pursuits and pluralistic government. The communist countries are dedicated to the idea of human betterment through the central control of the state and monolithic government. (These are over simplified statements of ideology but sufficient for purposes of example here.) Each pursues its own course vigorously in contest and competition with the other for control of the allegiance of other nations. And each can point to and document the flaws in the other's system. The flaws consist of the inability of each to achieve completely its own objectives and of not meeting certain objectives that the other holds most dear.

Where is the fear in this situation?

It is that the one will gain sufficient adherents and allegiance to overwhelm the influence of the other in significant regions of the world and thereby gain power to threaten the other's ideological and economic existence. Pointing to flaws in the other's execution of its ideology plus proclaiming the superiority of one's own ideology is a tactic in the struggle for power and control by each. Presumably the unaligned are persuaded by such tactics (a very doubtful presumption, by the way). Claims of "moral or social superiority" are continuously used by both sides as arguments. Such arguments are demonstrably reducible to personal bias and belief and have no basis in fact. They only superficially address the true problems of the contemporary era. Such claims now serve only to feed unrealizable expectations and exacerbate irrational fears rather than address the real problem of contemporary human need. The old game permitted one to claim the absolute of "rightness" on one's side. The new rules view that as being an absurd artifact of archaic thinking.

What are the rational fears involved in the ideological contest? Certainly they stem from somewhat different sources than that of MAD by nuclear exchange. They have to do with the basic needs of people in carrying out their respective and collective lives. In order to answer the question in a manner consistent with the previous treatment of military issues, we must develop a problem statement based upon the socio-political needs of people.

The socio-political needs of individuals may be characterized in terms of security, freedom, and control. Let's call this grouping SFC. Security labels the individual need for both physical and economic security. Freedom means the need for individuals to be able to make and execute choices for the significant aspects of their lives. Control addresses the need to be able to assure that the other two needs are met. Rational fears arise in individuals when the needs cannot be met in appropriate measure. Clearly the particular needs are not identical among individuals nor even constant within a given individual. SFC is a dynamic grouping of needs with trade-offs between the specific needs continuously taking place.

The needs of society are the aggregation of needs of the individuals within the society. These in principle are measurable as the means of individual measurement of needs. The aggregate social need will change much more slowly than the dynamic changes within individuals. A social organization is adopted in order to better orchestrate satisfaction of individual needs of all members of the society. (At least the foregoing is the avowed purpose of modern socio-political organizations but has not historically been the case nor is it necessarily the case today. The key word is "all." By substituting "certain members" for "all" we can generalize to all socio-political organizations.)

The definition of an organized society requires that some of the control needs to be yielded by the individual to higher levels of society in order to better satisfy the aggregate needs of the individuals within the society. Otherwise it is not an organized society, but anarchy. In exchange for yielding up a measure of control to society, the individual expects in return commensurate satisfaction of the other needs. Clearly certain organizational structures permit satisfying some needs better than others. A population with a high aggregate need for security and little ambition for personal freedom will tend toward socialist forms of organization as long as the promised security is delivered (which is likely not to be over significant periods of time). Populations with a high need for freedom and a desire for little control will tend to choose toward the anarchy end of the spectrum of possible organizations—until chaos proves untenable (which is also unsatisfactory over significant periods of time). We are assuming that people, through their control need, have an interest in choosing their form of social structure even though they may not have the freedom to do so. There has not yet been invented (and probably cannot be) the perfect social structure that can totally satisfy all the needs of all the people all the time. And certainly a totally static social organization, however well designed to satisfy initial conditions, is sure to eventually reach the point in a dynamic world where none of the needs of any of the people can be satisfied.

The foregoing can be developed at much greater length, but it is not necessary for the issues of this discussion. The idea we need to develop from the above is: 1) that rational fear arises when needs cannot be or are not being satisfied within the existing socio-political organization. Fear is a basic natural consequence of unsatisfied human needs; and 2) Fears also arise if an unsatisfactory socio-political organization threatens to dominate one's own system which one deems to be more satisfactory. The entire structure of SFC "need satisfaction" is threatened in this latter case. The first fear (and the attendant anger, which is a frequent consequence of fear) is normally considered an internal affair between the individuals and the managers of their social structure. But the superpowers routinely interfere in each other's internal affairs and those of other nations in promoting their own ideology. The second fear takes us back to the original problem between defensive and aggressive type societies but this time without the nuclear-space-technology issue dominating the arguments. The ideological conflict has its roots in the way socio-political structure is organized to satisfy human needs. But it is exacerbated by each society attempting to proselytize the world on behalf of its own ideology. It is in this context that the support role of conventional military forces as an adjunct of national policy is evident.

The superpower ideological conflict can be discussed in terms of the threat to or deprivation of the SFC needs defined in the foregoing. Each ideology would, and does, claim that its ideology best satisfies the needs of its own society. But each goes further to claim that its ideology also would best satisfy the needs of the population of the other society, indeed of all other peoples in the world. And proselytizing, of course, is where the difficulties commence. In the real world of the superpowers and playing under the traditional rules, the claim is made, or at least implied by the rhetoric, that "my socio-political system will best satisfy the world's SFC needs." Whether the claim of either could be measured against the world aggregate SFC need and in fact be substantiated might be interesting. It is not the validity of either claim that is of interest in this discussion. It is how the conflict between the claims can be best managed to achieve world stability that is of interest here.

It would be totally untrue to believe that either superpower has no concern for the satisfaction of the world's aggregate SFC needs. But it would be naive to believe that either has significant concern for world SFC needs much beyond those that affect the interests of its own population. Thus there is, reasonably, some small but finite amount of external altruistic interest by the superpowers' political management in satisfying the world's SFC needs. The problems arise because the rhetoric of both sides promises the world a lot, creates expectations, and seldom can deliver what it promises to the unaligned peoples. In fact, it is unlikely that either system can deliver what it promises in the long term by playing their old win-lose rules of the game. The superpowers play toward win-lose outcomes with each other, but that does not satisfy the SFC needs of the world community whose allegiance is sought and to whom promises are made.

5.4.5—HUMAN NEEDS AND FEARS

All strife, conflict and terrorism arise from unfulfilled human need and greed, fueled by a healthy dose of expectation and usually coupled with ignorance.

— Dr. Edgar Mitchell

The SFC (Security Freedom and Control) needs of the emerging nations of the world, the third world nations, are progressively getting further behind in satisfaction because of high population growth rates coupled with unstable governments that are unable or unwilling to address the problem in proper measure. Thus the rational fear level of the populations increases and will continue to increase until their pressing and legitimate SFC needs can be satisfied. Such SFC needs cannot be satisfied in the face of political instability, nor could they be satisfied even were the total combined assets of the industrial world equally distributed across the third world. They can only admit of solution with socio-political stability over significant periods of time if indeed they can be satisfied at all. SFC needs can only be satisfied when a societies' production-consumption activities internally produce as much as they consume. Long term stability is the only way in which this can happen. But with the population growth rates in the less developed nations threatening to double the world's population during the next 50 years, there is little likelihood of global stability unless the industrial world helps create it.

The third world nations and the frightened and angry people within them know little and care less about the new rules of the game. They are not interested in the superpower struggle except as they can use it for sat-isfying their own needs. They universally want new and modern weapons to better play the old game of power and military escalation in order to take what they want and believe to be theirs. There is no concept that the traditional "win-lose" strategies degenerate to "lose-lose" as military tech-nologies escalate. And the superpowers oblige by providing the weapons that perpetuate the instability.

Because the superpowers, indeed the entire world, has been playing by the rules designed toward "win-lose" outcomes, the SFC needs in 'the third world continue to increase. One superpower provides military and financial assistance to the group in power, and the other provides military and financial assistance to the dissidents. But the SFC needs which serve as the driving force for fear, anger and armed conflict do not get satisfied because the political situation cannot become stable. Each superpower is prepared to help dissidents topple an unfavored regime if they have chance of success or help sow general confusion if they don't. This situ-ation has prevailed in one form or another throughout recorded history but has intensified since World War II as the world has polarized into East-West alignments.

The danger is that the world population explosion that exacerbates the SFC needs of the third world countries is driving the institutions that traditionally help satisfy those needs into the same cul-de-sac that military technology has been driven by SDI. In other words, the old rules of the game that produced win-lose solutions as an accepted norm of human behavior, begin to produce lose-lose solutions under modern conditions. SDI drives military technology into a lose-lose condition. The population explosion, particularly in the third world, is driving the normal win-lose strategies of socio-political decision making into similar lose-lose situ-ations. If indeed the third world fears because of unfulfilled SFC needs continues to increase, both of the superpowers, in fact, the entire in-dustrial world will be caught in a global lose-lose scenario in which there can be no winners.

With that foreboding idea, lets attempt to construct an amendment to the problem statement of "freedom from fear of attack" that might better manage superpower conflict with its military policies and thereby lead toward true satisfaction of aggregate SFC needs which would, in turn, lead toward global peace. It might seem that "freedom from fear" would be sufficient. That sounds exceedingly simple, probably too simple. Likely it is a goal that is unattainable in the foreseeable future, too idealistic, and probably unnecessary in any all-encompassing sense. But we can use the general form: "freedom from fear of _____" and allow any rational fear to be inserted. Again, a rational fear is one associated with an undesirable event which has a measurable and verifiable probability of occurrence. The foregoing, with strategies aligned toward win-win solutions, might assist in preventing the current win-lose strategies from degenerating into lose-lose outcomes. It would require, however, East-West cooperation rather than altercation.

There are certainly other approaches to defining policy statements at the personal, national, and international levels. However, such an approach as suggested in this section has the benefit of recognizing that human needs, and particularly human fears, drive a huge portion of human activity individually and societally. Certainly fear drives military activity even though we may be loath to admit it. By quantifying undesirable events in a form manageable with the mathematics of statistical probability then giving them a priority for action, we combine scientific principles of statistical analysis with the most important of all motivators, human need. Human needs and the fears that result from deprivation of need are easily and quickly related in this fashion. Fear is directly related to the probability of an undesirable event of need deprivation occurring.

By deriving policy statements along the lines suggested in this section for managing superpower conflict along a path toward win-win solutions and thus toward global peace, positive results can be accomplished. In the following section we will point out positive cooperative programs already in existence and propose some that would follow such strategies and lead to the point where concepts such as SDI would not be needed. None of these ideas are going to eliminate human conflict, for that is necessary to stimulate creativity and progress. They will, however, lead to better ways of managing the conflict short of armed confrontations.

5.5—COOPERATION—SANITY DOES EXIST!

We need new perspectives and approaches to technology and management.

— General Abrahamson, SDI Director

Human beings have been developing the space frontier for the past three decades, internationally. What began as a space race evolved in the 1970s into many cooperative efforts and joint endeavors. There have been over 200 humans from 19 nations launched into space aboard U.S. and Soviet spacecraft. There are as yet no weapons in space, and there has been co-operation! There's a successful history of joint U. S./U. S. S. R. efforts on Apollo-Soyuz; on space science, on other science, on cultural exchange, with an impressive list of long-range commitments and new proposals both in space and for Earth bound projects date which date to the early 1970s. These have included agreements with the following agencies of the U.S. government: (see Appendix A3 for details)

- National Academy of Sciences
- Department of Agriculture
- Department of Transportation
- Department of Housing and Urban Development
- Department of Energy
- Department of Commerce
- National Aeronautics and Space Administration

– National Science Foundation
– Department of Health and Human Services
– Department of the Interior

In addition, there has been increasing private sector exchange in the form of commercial activity, cultural exchange and tourism which resulted in excess of 45,000 Russian visas being issued to Americans in 1985. Soviets hold in excess of 53,000 patents in the U.S. and have earned in excess of $80 million in royalties from American business.

The breakdown of suspicion and fear through such exchanges and cooperative projects has been progressing for almost two decades, with setbacks admittedly, but with progress and success prevailing in all areas By continuing cooperatively to participate in research and development of the space frontier as well as more down-to-Earth projects, we can produce satisfying results. SDI can only be seen as a threat to the momentum already gained in East-West cooperative ventures which should tend to eventually "normalize" relations. And we must ask: "Which avenue of development seems to offer the greatest hope of long-term security?"

Space age technologies and services can provide direct benefits and solutions to urgent human and environmental problems on a global basis. And can do so in ways superior to any other technologies. For example, images from space have created a new environmental consciousness. We can see the polluted waters in the oceans, and guide fishermen (and fisherwomen) out to the clean waters, globally. We can track the patterns of forest and crop damage from acid rain back to their sources. We can monitor the polar regions minutely for their weather-making changes and for atmospheric ozone stability.

For military planners, space systems have contributed thus far to stability with satellite sensors only and without the use of space weapons. We can expand space systems to better verify arms agreements and to monitor potential areas of conflict. Such surveillance systems can also be used cooperatively to monitor drug related agriculture to better control that social malignancy. These types of problems, and they are only examples, know no national boundaries and require international cooperation to achieve satisfactory solutions.

Telephone, television, and data transfer via communications satellites are increasingly in routine use and have transformed the global economy. No longer does any small region of the world need to be isolated but can become a part of and contribute to a competitive world economy which wants new products and markets. Such commerce begins with inexpensive and fast communications which satellites can provide. Industry representatives already see space as a medium they can use to develop challenging new markets, finding profit and an abundance of jobs. Space flight provides us with a unique new medium for endeavor and which has produced new perspectives; a pragmatic new approach to solving problems of Earth that previously had no means of solution. And, as yet, unsullied by weapons to threaten this development.

Because the vast majority of humans on the planet today will never be afforded the experience of the few astronauts and cosmonauts, to observe firsthand that almost mystical vision of Earth as a tiny speck of dust in a great black void, you must try, with imagination, to look deep inside yourself, to see with their eyes. To see that space by its nature requires cooperation; it is boundary-less, boundless, and fraught with infinite potential—or danger if wrongly used.

In this era, the very existence of life on Earth is in jeopardy. All nations have professed a desire to reduce and finally eliminate the threat of massive nuclear arsenals. That threat and the fear which attends it diverts attention from the other pressing issues of this Earth. As astronaut Russell Schweickart described, "You realize that on that small spot, that little blue and white thing, is everything that means anything to you—all of history and music and poetry and art and death and birth and love, tears, joy, games, all of it on that little blue spot out there that you can cover with your thumb. And you realize from that perspective that you've changed, that there's something new there, that the relationship is no longer what it was."

Cooperation in the space frontier is optimistic, idealistic, but realistic. Cooperation is what is really going on—and must always go on—in space! It can make human dreams for security, prosperity, peace and abundance on Earth possible in ways hither-to-impossible if we direct our energies in positive directions rather than the negative directions of fear, mistrust and confrontation. From a distance, Earth looks peaceful, united and of common purpose. Why does it not seem so at close range?

5.5.1—BUT WHAT ABOUT THE RUSSIANS?

The U.S. and the Soviets are changing. This change must be away from confrontation toward cooperation. This change must start now. By taking a leading role in promoting this vision, the U.S. will manifest its basic high standard, character, and principles. Just as all our pressing problems have been created by human beings, so will the combined creativity of human beings meet these challenges and build a better world.

— Major General Jack Kidd, U.S. Air Force, Retired

General Kidd promotes an idea he has named "STARLIGHT." It is a plan for a new national strategy, Strategic Cooperation Initiative. There is strong evidence that most active military space planners privately prefer this idea, as do Soviet representatives. So, what is holding us back? Our fear? Our lack of trust? Our perceptions of each other's motivation? The things Soviets do or the way they are? Or just an entrenched adversarial tradition?

We don't need to completely trust in each other to cooperate in space development. That can come later as it evolves and when it is warranted. We have continually agreed to major lists of joint ventures and produced them. The Soviet entry into the market for commercial space launches in 1986–7 indicates a new way of "playing the game" (Appendix A3).

The Soviets have agreed to eliminate nuclear testing and weapons, to ban weapons from space and to cooperate in space ventures, if the West will do likewise. It is up to the American people to elect leaders who want the cooperative vision rather than a confrontational one. It is time to put aside the ancient quarrels. The era demands that we must. It is only a matter of political will. It is about time we support a Soviet proposal if it makes sense rather than viewing it with unrelenting suspicion simply because of its origin; to build on the sensible and feasible and let that lead to softening of attitudes in other areas. The vision of the old Bolshevik, Nikita Khrushchev, banging his shoe on the table at the United Nations is three decades behind us. Let's leave it there and let more moderate voices prevail. The West has and has had its share of radical rhetoric as well. Neither offers much to be proud of. The West has not been "buried" as promised, nor will it be. Quite the contrary. But neither will the communist philosophy be eradicated from the mind of the unaligned poor and indigent by force of arms or by causing failed expectations. Their need can only be exacerbated in that fashion. Only by positive experience, accurate knowledge and satisfaction of legitimate needs can political structure hope to endure in the modern era. All other means and ends are increasingly seen as amoral, self-serving and corrupt.

It is time to act upon the truth about the common global need and common best interests. Based on that truth, we can continue productive, on-going ventures. We can undertake the already-proposed joint space ventures, as well as Earth-bound ventures that solve problems for us all rather than one, SDI, which creates new problems. Cooperative ventures and economic necessity had already begun to soften attitudes; to replace the need for space-based weapons and nuclear confrontation but that fact is not yet politically nor generally acknowledged in the zeal to create confrontation with the "forces of evil."

5.5.2—SIMULTANEOUS REALITIES

You shall know the truth and the truth shall make you mad.

— Aldous Huxley

Many realities exist simultaneously, as many as there are humans to interpret what reality is. For it is the individual interpretation of reality, or "world view," not objective reality itself, that guides human decisions. Objective reality is but an elusive idealization to be sought and approximated but never really known. It is why the authors use the words "reality" and "game" almost interchangeably. Because the mind interprets its own version of reality, we continuously rediscover our prejudices and reinforce them. But we can change our minds. We can choose. Confrontation or cooperation is a matter of choice.

The weapons game and the cooperation game are being played simultaneously. Which will prevail? One stems from a reality based on fear, uncertainty, secrets, paranoia, isolation, and confrontation; the other is based on developing trust, recognition of oneness and interconnectedness, cooperation and collaboration. And maybe, dare we say it, love, respect and caring. Both are methods for dealing with human conflict, but one method assigns total responsibility for error to the opponent, the other requires acceptance of personal responsibility and self-examination as well to achieve conflict resolution.

For the issues of this book, time is of the essence. The decision to move humans into space was made decades ago. The time has come to reaffirm if we shall continue to do it in peace or to make space another arena for war, the latter likely being an irrevocable decision. In this period of decision, between a more positive vision of reality and a traditional but negative one, the future will be determined for generations to come.

We had been building and still can continue momentum for cooperative space development. Any resources allocated to the SDI for space weapons research is a diversion from this positive vision. SDI resources can be redirected, however, and directly applied to non-weapons cooperative research. The same institutions and expertise would still be necessary, just the emphasis changed.

5.5.3—WIN-WIN WITH SPACE COOPERATION

We must transform the dynamics of the world power struggle from the nuclear arms race to a creative contest to harness man's genius for the purpose of making peace and prosperity a reality for all.

—Vincent J. Intondi

The challenge is to be aware of and to accustom ourselves to the reality that was already building both in international relations and in the space frontier: to opt for continuing that reality: cooperation. Ironically, even with dedicated attempts to avoid cooperation and trust, both sides must still consider each other's responses and the need for cooperation to play the war game short of going to war. In other words, some form of cooperation must take place even with the SDI. Unilateral deployment of SDI space components cannot take place without an unacceptably high risk of war. Soviet acquiescence would be necessary for deployment in order not to begin the shooting.

Actually, all realities call for some form of trust, even if it is only trust in the opponent's untrustworthiness. Building trust via communication and cooperation, however, helps reduce fear rather than increasing it. Win-win relations already exist. Our task is to understand this and build upon them. Over 300 satellites orbit every nation on the Earth. They perform an immense variety of functions. The U.S. and Soviet militaries have paved the way for both military and civilian ventures in space.

223

Industries have built communications satellites that transport two-thirds of all international telephone and television transmissions. This satellite technology, including live two-way audio-visual teleconferences, has added stability to U.S. and U.S.S.R. relations by providing an increasingly used medium for communication. There have been numerous live TV exchanges between Soviet and American audiences in this manner under a private project named Track II. Direct citizen diplomacy via such satellite teleconferencing has enjoyed many successes already.

Since the dawn of the Space Age, the U.S. has signed over 1000 agreements with 100 nations in cooperative space activities. The U.S. helped form the 112 nation International Telecommunications Satellite Organization (INTELSAT) and the sister maritime agency, INMARSAT, which also counts the Soviet Union as a member. Both nations initiated the United Nations Committee on the Peaceful Uses of Outer Space. These major international bodies have been very successful in providing services and anticipating potential conflicts.

In the past, the majority of joint space projects were carried on between allies of either the U.S. or the U.S.S.R. Now, there is more direct cooperation across opposing political boundaries. Satellites were the first space technologies that proved our natural, international interdependence, and interconnectedness. And they serve this function admirably.

5.5.4—ALTERNATIVES FOR NATIONAL AND INTERNATIONAL SECURITY

The possibility of reclaiming a peaceful future of humankind and a safe environment on Earth is intimately linked with curbing our violence towards each other.

— Carol Rosin

The common and fundamental priority of all world citizens is an adequate means of survival. This has been expressed herein in terms of the SFC needs: security, freedom and control. These needs must lie at the heart of any plan for achieving global stability and peace. International space cooperation offers powerful technologies to address these fundamental needs. The latent opportunities to better satisfy SFC needs must be identified and expanded, for these needs are primary and underlie all other issues which lead to conflict.

We have commonly thought of security as that sense of safety provided by weapons: If I felt rich in weapons, I felt powerful and free from anxiety. Indeed, the very word secure, as derived from the Latin, means "carefree." But how carefree can I feel sheltered under an umbrella of weapons the magnitude of which, if used, can at any moment obliterate my entire world? This awareness prompts a broader definition of personal and national security, which must include global security. None can truly be secure until all are secure. Horizons have expanded; so must our thinking. We must come to realize that we cannot exist independently of each other. Perhaps we are not yet near the point in history at which we can jettison the concept of an "enemy." But we are, in fact, proving to ourselves and each other, year in and year out, that we can redefine how we consort with our "enemy." What space flight has done, almost in spite of ourselves, is to provide us the necessity for a new definition and an imperative to use it. In actuality, security in the modern era lies in information shaped into new knowledge and new action—not in steel shaped into new weapons with old actions.

For perhaps the first time since World War II, the distribution of political power among the world's major nations is more determined by economic means and potential than by military prowess. Such relationships are hard to measure, but, nevertheless, economic competition among the world's powers is a more potent force than traditional military competition for shifting the balance of power. The evidence for this assertion comes from many sources. Clearly, the U.S., Western Europe and Japan prefer to compete, with each other and the rest of the world, on economic rather than military terms. In fact, "competitiveness," for better or worse, has become the new "watchword" in the U.S. Even the Chinese have recently shown initiatives in revitalizing their economy and in putting high priority on economic growth through competitive entrepreneurship. We see few indications of their desire for near-term military driven expansion. Perhaps because there are no new land areas for conquest, as in previous centuries, military driven expansion can be seen as an archaic notion, no longer viable.

Finally, the Soviet Union, under Gorbachev's leadership, clearly wants a relaxation of military tension to enable them to focus on economic development and glasnost (openness). They want and need this badly because the West has already proved that it creates the wealth that societies need. Unfortunately, just at the moment the East had begun to move in the positive directions that the West had been wanting, the U.S. government, under radical leadership, began its retreat into secrecy, hostile aggressiveness, covert maneuver and deception. Retreating into a hypocritical self-righteousness unbecoming to the leadership of a major world power. But such an approach is unproductive, destabilizing and cannot long endure. There is a significant danger, however, that contemporary Western attitudes, if sustained, can drive the fledgling initiatives of the East back into the old and dangerous policies.

The key to continuation of the movement toward cooperation and openness is verification of treaties and arms limitation agreements.

Today's verification techniques are better than ever. With satellites, seismographs, and other sophisticated technologies, we observe every launch and detect every nuclear test that takes place on this planet. But remote technical means alone are not enough. They lack the human interaction that alone dispels fear and distrust between opponents.

The Soviet Union has already agreed to and permitted seismic monitoring of nuclear test sites by the National Resource Defense Council (NRDC), and in principal has offered inspection at space launch sites. On-site verification at space launches supplements and validates existing technological techniques. Whether we realize it or not, the nature of our relationship had already begun to change.

The remainder of the world is not without voice in this regard. The World Parliamentarians have been working with the Five Continent Peace Initiative, (the heads of state of Greece, Mexico, Argentina, India, Sweden, and Tanzania), to form a neutral international team of experts to verify agreements for arms reduction and elimination. Such independent thinking from the involved bystanders to superpower conflict is vital.

5.5.4.1 THE TECHNOLOGY

We now possess the power and knowledge to control Earth, but do we yet have the wisdom and sense of destiny necessary to control the technological genie we have set free?

— Dr. Edgar Mitchell

Not only is the number of nations with satellite monitoring capabilities increasing rapidly, but the technology to conduct such activities is being revolutionized. Active positioning technology, like radar, and passive localizing technologies can be used to track nuclear fuel to help prevent nuclear proliferation. Sealed containers with transmitters can be tracked from space. Hidden supplies of nuclear fuel can be detected. Such technologies can also be used in conjunction with other non-weapons military space technology, such as signal monitoring and communication satellites, to help eradicate the global threat of terrorism by allowing increasingly sophisticated intelligence, coordination, and communication.

A problem is that no one nation alone can afford to develop and use all these systems individually, but the need is a need for all. We are, therefore, impelled toward cooperative endeavors. These capabilities make it potentially impossible for surprise attack to occur and make covert testing and development of advanced weapons very difficult and risky. In short, when such systems are deployed, they will make major contributions to the security of nations having access to the data they collect.

This is just one example of how cooperative space technologies, in place of weapons systems, can be deployed. The information gathering gives ample time for planning armed response were it ever to actually be necessary. It is clear that those industries that are interested in obtaining a slice of defense budgets should want to work on such non-weapons development. Expanded international military use of sensing satellites to monitor troop movements and to enhance on-site inspection (on Earth and in space) can help to enforce verifiable agreements.

Proposal: Imagine the militaries of the U.S., U.S.S.R. and other rations working together on joint space command posts, on Earth and in space. The military has previously used space for communications, surveillance, and navigation independently. And without space-based weapons! Unlikely as such a scenario may seem at this time, it is entirely feasible in the very near term based on evolving concepts of cooperation. My experience suggests that most enlightened members of the military establishment would welcome a directive from their Commander in Chief ordering them to the field of cooperation rather than the field of battle. Working on cooperative ventures in space research and development, whether civilian or military, give them the opportunity to provide true service and security. The same careers and challenges would be available but without the need for weapons.

Even in the field of military strategy, cooperation between the U.S. and U.S.S.R. is inevitable. If the Soviets accepted sharing the SDI system (as initially proposed), then obvious collaboration would be needed—but then SDI would not be needed. If the Soviets refused to accept SDI collaboration, the U.S. would still need Soviet cooperation to limit offensive missiles or even to deploy SDI without beginning war.

Now let's talk about some of our contemporary real enemies rather than the traditional ones.

5.5.5—ENVIRONMENTAL SECURITY

Civilization needs to give a great deal of thought to exactly what it's doing to itself.

—Dr. Edgar Mitchell

Environmental dangers threaten the security of our nation and our planet. We need to defend ourselves against these "enemies."

Examples: For every person in the U.S. there is one ton of toxic chemical waste. This has made the planet a biochemical time bomb. The damage caused by acid rain threatens to destroy 300 lakes in the northeast over the next fifty years, according to the Environmental Protection Agency. Clearly, the world cannot continue to pollute the land, waters and atmosphere without causing epic problems for the environment. There is no easy solution for problems caused by burying nuclear waste. All realistic solutions are difficult and costly and should be cooperatively implemented. The problem is really more political than technical, however, since technical solutions are in hand. There are 40,000 or more pieces of visible space junk threatening astronauts and cosmonauts and expensive investments in space: a threat shared by all. Explosions in space increase that hazard.

From satellites we've seen that our ground water, rivers and oceans, are polluted and threaten the survival of animal species, including ourselves.

Proposal: Let us set up joint space laboratories and industries to perform research that will provide us with technology and information to help us clean up the environments of Earth and space, research renewable energy resources from the sun, wind, and water, and use our cooperative satellites for observing the ozone layer (nearing destruction if industrial damage continues), and perhaps to defend Earth from approaching asteroids or large meteors. International cooperation is necessary. We can only arrive at solutions to these common problems by further pooling our resources and developing joint goal-oriented projects.

5.5.6—HUMAN NEEDS

The technology of civilization has totally outstripped its morality.
— Dr. Edgar Mitchell

Starvation, epidemics and ignorance are related and common "enemies." We have the resources, means of transportation, and surplus food potential to feed the entire world, yet millions of people starve. We have health and hygiene techniques available, yet epidemics spread. We have educational potential but wide illiteracy.

Proposal: Let us direct space satellites to monitor the migration of animals, to manage food and water resources, to conduct inventories of agriculture and forests, and to educate us as we work together to see that human suffering ends. Let us set up international laboratories to produce new products requiring the special environment of space for manufacture. Let us extend the training of doctors and nurses around the world via satellite and video conferencing facilities to bring experts together for problem solving. Let us build habitats in space, where experts can extend their experience and knowledge through that unique environment to discover new techniques that can be applied to Earth. Educational facilities can be made available through satellite systems to 60% of the world's illiterate. Let us expand the COSPAS-SARSAT[6] search and rescue satellite

---

6   COSPASS-SARSAT is a satellite-based monitoring system that detects and locates emergency beacons. Professional operators then notify search-and-rescue (SAR) authorities. The beacons comply with internationally agreed standards for radio communication and identification of beacon owners.

system, which has already saved hundreds of lives locating downed planes, or lost ships, even a lost child on a mountainside.

This age of space technology affords us possibilities of information gathering and organization never before dreamed of that can help secure life on Earth. By communicating, collaborating, and cooperating, we can better understand each other and modify our systems for the betterment of life. Through this understanding, we learn to fear less and to build trust. Rather than smarter, faster, more accurate weapons in space, we can use space to develop more educated and experienced people, faster, to safeguard our planet.

5.5.7—ECONOMIC SECURITY

It was pretty obvious, from Hiroshima in WWII, that if we ever really began using nuclear weapons, civilization as we knew it was at an end.

— Dr. Edgar Mitchell, 2005

All of the grand schemes for cooperative human betterment require financial support, thus a vital world economy. Development of the space frontier has huge significance for the world economy. The U.S. has been a leader in plans for commercial space ventures. But the U.S. is hardly alone. The first official International Business in Space Conference was held in 1978. At first, the Eastern bloc did not participate. Now, anyone can return home from talking with an Eastern spacefaring nation with a plate full of joint economic ventures.

Today, we all are dependent on a world economy that is not subject to any one country's control. Market forces and human ingenuity in using them drive the activity. Today, private and low-level negotiations, with public but little top-level U.S. support, are pushing for increased international exchange and cooperation in response to these forces. It results in better security and social benefits as well as economic gain. With proper priority on such efforts these would bear more fruit. Fortunately, the same basic technologies needed for space weapons can find non-weapons space applications and thus strengthen this private "push" toward mutual security, economic return and jobs. "Forging Missiles into Spacecraft" as scholar Daniel Deudney puts it. We are experiencing major breakthroughs in awareness about alternative approaches at the private level, as well as breakthroughs in space technology, but have experienced a regression in thinking at governmental levels on dealing with the Eastern bloc.

Hundreds of industries and entrepreneurs have entered the space marketplace. New products and services are being researched. Economic factors are drivers and are now based on international, not just national interests. From national and corporate treasuries to wage earners pocketbooks, money is going to be made from space-based investments. Short or long term, this is inevitable.

The U.S. has lost its leadership role in several important ways as the U.S. government has de-emphasized its commercial priorities in space to emphasize military priorities. Other nations have filled in and undertaken successful competition and they will be the winners in know-how, social and economic return should this continue. France, China, and the U.S.S.R. are offering to orbit commercial satellites planned for the Shuttle.

Proposal: Let us re-emphasize commercial and scientific space endeavors rather than weapons-oriented endeavors. The emphasis at the moment is exacerbating a problem that was already well on the way to solution.

5.5.8—EXTERNAL THREAT

An asteroid can literally destroy 80 or 90 percent of the species that are alive on Earth. These are big events. I mean, this is called extinction.

— Apollo 9 Astronaut Rusty Schweickart
Chairman, B612 Foundation and Asteroid Institute

There is one eventuality that the world must sooner or later face together. It could become larger than any other faced in historic times and which had no solutions before the present. It is the threat of collision with asteroids, comets and large meteors, even perhaps, "black holes." The geologic history of both the Earth and the Moon attests to the many large impacts that have occurred in the past. Though now with decreasing frequency, they have a finite probability of still occurring. Preventing such a collision could represent an opportunity for the final and only productive use of nuclear explosives in the future. It could require a small but significant portion of existing nuclear capability coupled with the best launch and guidance technology that can be devised. Such solar objects would need to be intercepted far out in space and diverted or destroyed by nuclear impact. Clearly such effort is a prime contender for U.S.–Soviet cooperation—and an appropriate graveyard for nuclear weapons technology.

5.6—FINALE

We have placed our civilization and our species in jeopardy. Fortunately, it is not yet too late. We can safeguard the planetary civilization and the human family if we so choose. There is no more important or more urgent issue.

— Carl Sagan, *Parade* magazine, October 30, 1983

The illusory security we yearned for from space-based weapons can never equal the real safety achievable by cooperative space endeavors and cooperation in general. This is the approach that can put meaning into the changed concept of "defense." Neither offensive nor defensive space-based weapons defend us, and they divert us from the life enhancing thrust of cooperation. SDI and space weapons research and development must stop. Fundamental scientific space related development must continue and expand. And the latter does not justify the former. It is not a means to an end, the means and the ends must each justify themselves.

The aerospace-military industry, labs, and universities would have little trouble converting away from space weapons components, as the SDI program has not yet developed an irreversible economic or political momentum. In fact, the process for developing the space frontier can continue as is, minus the intention to develop space weapons because the same basic technologies are needed. Lasers are needed for communications, for example, if not for weapons. Compromises and mere reductions in numbers of weapons may be a step, but alone do not end the arms race or the old adversarial game. Provided we reassert a climate of cooperation, it is not unrealistic to envision and achieve an actual end to the superpower arms race and thereby set a new example for emerging nations to follow.

The existing political and institutional alignments stemming from tradition, secrecy, suspicion and fear can be overcome with collaboration and cooperation in a spirit of openness and satisfaction of common need. Deterrence by promise of cooperation rather than threat of retaliation should be the new thrust for strategic policy! Fear is a common human enemy. We need proceed with positive programs of the magnitude of the SDI to enhance common wellbeing rather than to deterrence through nuclear arms.

If humankind truly desires peace and prosperity rather than Armageddon, now is the time to choose it.

– Dr. Edgar Mitchell, May 1987

APPENDICES

A1—NOTES AND REFERENCES

1.   John Tirman, editor, Union of Concerned Scientists. *Empty Promise* (Boston, MA: Beacon Press, 1986), pgs. 203–210.

2.   Nicolaas Bloembergen et al. "Science and Technology of Directed Energy Weapons," Report to the American Physical Society, April 23 (1987).

3.   Donella H. Meadows, et al. Limits to Growth (Falls Church, VA: Potomac Associates, 1972).

4.   Alvin Tofler. *Future Shock* (New York, NY: Bantam, 1984).

5.   Aurelio Peccei. *The Chasm Ahead* (London: The Macmillan Co., Collier-Macmillan, 1969).

6.   Ernst Schumacher. *Small is Beautiful* (London: Blond & Briggs, 1973).

7.   John Naisbet. *Megatrends: Ten New Directions Transforming Our Lives* (New York, NY: Warner Books, 1982).

8.   Hawken, Paul, James Ogilvy, Peter Schwartz. *Seven Tomorrows* (New York, NY: Bantam Books 1982).

9.   Ilya Prigogine and Isabelle Stengers. *Order Out of Chaos* (New York, NY: Bantam Books, 1984).

10. Paul Faure. *Alexandre* (Paris: Librairie Artheme Fayard, 1985).

11. Napoleon III. *The History of Julius Ceasar* (1865).

13. Richard Nixon. *Real Peace* (Boston: Little Brown and Co., 1984).

13. Richard P. Turco, Carl Sagan et al. "Nuclear Winter: Global Consequences of Multiple Nuclear Explosions." *Science*. Vol. 222, Issue 4630 (1983) pp. 1283–1292.

14. Eric Stubbs, Rosy Nimroody, Ed. "Star Wars: The Soviet Program," in *Star Wars: The Economic Fallout* (Cambridge, MA: Ballinger Press, 1986).

15. Robert M. Bowman. *Star Wars: Defense or Death Star?* (Potomac, MD: Institute for Space and Security Studies,1985).

16. Yevgeni Velikhov, Roald Sagdeev, Andrei Kokoshin, translated by Alexander Repyev. *Weaponry in Space: the Dilemma of Security* (Moscow: Mir Publishers, 1986).

17. DeGrasse Hartung, et al. *The Strategic Defense Initiative: Costs, Contractors & Consequences* (New York, NY: Council on Economic Priorities, 1985).

18. John von Neumann and Oskar Morgenstern. *Theory of Games* (Princeton, NJ: Princeton University Press, 1944).

19. Michael Kidron and Ronald Segal. *The State of the World Atlas* (New York, NY: Simon and Schuster, 1981, 1983).

20. Michael Kidron and Ronald Segal. *The New State of the World Atlas* (New York, NY: Simon and Schuster, 1983).

21. Thelma Liesner. *Economic Statistics 1900-1983* (London: The Economist Publications Ltd., 1985).

22. Michael Kridon and Dan Smith. *The War Atlas: Armed Conflict—Armed Peace* (New York, NY: Simon and Schuster, 1983).

A2—ANTI-BALLISTIC MISSILE TREATY

This appendix contains the wording of the Anti-Ballistic Missile Treaty of 1972, an additional protocol to the Treaty signed in 1976, plus a number of "common understandings" initialed by the heads of the delegations in order to clarify certain terms and concepts used within the Treaty. It is recommended by the authors of this book that the reader read the texts of these documents in order to be familiar with items currently under debate. The text is quite direct and uncomplicated. The authors have provided commentary at the end of this appendix regarding current U.S. efforts to "reinterpret" and/or abrogate the Treaty.

ANTI-BALLISTIC MISSILE TREATY OF 1972

Treaty between the United States of America and the Union of Soviet Socialist Republics on the Limitation of Anti-Ballistic Missile Systems

Signed at Moscow May 26, 1972
Ratification advised by U.S. Senate August 3, 1972
Ratified by U.S. President September 30, 1972
Instruments of ratification exchanged October 3, 1972
Entered into force October 3, 1972

The United States of America and the Union of Soviet Socialist Republics, hereinafter referred to as the Parties,

Proceeding from the premise that nuclear war would have devastating consequences for all mankind,

Considering that effective measures to limit anti-ballistic missile systems would be a substantial factor in curbing the race in strategic offensive arms and would lead to a decrease in the risk of outbreak of war involving nuclear weapons,

Proceeding from the premise that the limitation of anti-ballistic missile systems, as well as certain agreed measures with respect to the limitation of strategic offensive arms, would contribute to the creation of more favorable conditions for further negotiations on limiting strategic arms,

Declaring their intention to achieve at the earliest possible date the cessation of the nuclear arms race and to take effective measures toward reductions in strategic arms, nuclear disarmament, and general and complete disarmament,

Desiring to contribute to the relaxation of international tension and the strengthening of trust between States.

Have agreed as follows:

Article I

1.   Each Party undertakes to limit anti-ballistic missile (ABM) systems and to adopt other measures in accordance with the provisions of this Treaty.
2.   Each Party undertakes not to deploy ABM systems for a defense of the territory of its country and not to provide a base for such a defense, and not to deploy ABM systems for defense of an individual region except as provided for in Article III of this Treaty.

Article II

1.   For the purpose of this Treaty an ABM system is a system to counter strategic ballistic missiles or their elements in flight trajectory, currently consisting of:
    (a)   ABM interceptor missiles, which are interceptor missile constructed and deployed for an ABM role, or of a type tested in an ABM mode;
    (b)   ABM launchers, which are launchers constructed and deployed for launching ABM interceptor missiles; and

(c)   ABM radars, which are radars constructed and deployed for an ABM role, or of a type tested in an ABM mode.

2. The ABM system components listed in paragraph 1 of this Article include those which are:
  (a)   operational;
  (b)   under construction;
  (c)   undergoing testing;
  (d)   undergoing overhaul, repair or conversion; or
  (e)   mothballed.

Article III

Each Party undertakes not to deploy ABM systems or their components except that:
  (a)   within one ABM system deployment are having a radius of one hundred and fifty kilometers and centered on the Party's national capital, a Party may deploy: (1) no more than one hundred ABM launchers and no more than one hundred ABM interceptor missiles at launch sites, and (2) ABM radars within no more than six ABM radar complexes, the area of each complex being circular and having a diameter of no more than three kilometers; and

  (b)   within one ABM system deployment area having a radius of one hundred and fifty kilometers and containing ICBM silo launchers, a party may deploy: (1) no more than one hundred ABM launchers and no more than one hundred ABM interceptor missiles at launch sites, (2) two large phased-array ABM radars comparable in potential to corresponding ABM radars operational or under construction on the date of signature of the Treaty in an ABM system deployment area containing ICBM silo launchers, and (3) no more than eighteen ABM radars each having a potential less than the potential of the smaller of the above-mentioned two large phased-array AMB radars.

Article IV

The limitations provided for in Article III shall not apply to ABM systems or their components used for development or testing and located within current or additionally agreed test ranges. Each Party may have no more than a total of fifteen ABM launchers at test ranges.

Article V

Each Party undertakes not to develop, test, or deploy ABM systems or components which are sea-based, air-based, space-based, or mobile-land based.

Each Party undertakes not to develop, test, or deploy ABM launchers for launching more than one ABM interceptor missile at a time from each launcher, not to modify deployed launchers to provide them with such a capability, not to develop, test, or deploy automatic or semi-automatic or other similar systems for rapid reload of ABM launchers.

Article VI

To enhance assurance of the effectiveness of the limitations on ABM systems and their components provided by the Treaty, each Party undertakes:

(a) not to give missiles, launchers, or radars, other than ABM interceptor missiles, ABM launchers, or ABM radars, capabilities to counter strategic ballistic missiles or their elements in flight trajectory, and not to test them in an ABM mode; and

(b) not to deploy in the future radars for early warning of strategic ballistic missile attack except at location along the periphery of its national territory and oriented outward.

Article VII

Subject to the provisions of the Treaty, modernization and replacement of ABM systems or their components may be carried out.

Article VIII

ABM systems or their components in excess of the numbers or outside the areas specified in the Treaty, as well as ABM systems or their components prohibited by this Treaty, shall be destroyed or dismantled under agreed procedures within the shortest possible agreed period of time.

Article IX

To assure the viability and effectiveness of this Treaty, each Party undertakes not to transfer to other States, and not to deploy outside its national territory, ABM systems or their components limited by this Treaty.

Article X

Each Party undertakes not to assume any international obligation which would conflict with this Treaty.

Article XI

The Parties undertake to continue active negotiations for limitation on strategic offensive arms.

Article XII

1. For the purpose of providing assurance of compliance with the provisions of this Treaty, each Party shall use national technical means of verification at its disposal in a manner consistent with generally recognized principles of international law.

2. Each Party undertakes not to interfere with the national technical means of verification of the other Party operating in accordance with paragraph l of this Article.

3. Each Party undertakes not to use deliberate concealment measures which impede verification by national technical means of compliance with the provisions of this Treaty. This obligation shall not require changes in current construction, assembly, conversion or overhaul practices.

Article XIII

1. To promote the objectives and implementation of the provision of this Treaty, the Parties shall establish promptly a Standing Consultative Commission, within the framework of which they will:

    (a)   consider questions concerning compliance with the obligation assumed and related situation which may be considered ambiguous;

(b) provide on a voluntary basis such information as either Party considers necessary to assure confidence in compliance with the obligations assumed;

(c) consider questions involving unintended interference with national technical means for verification;

(d) consider possible changes in the strategic situation which have bearing on the provision of this Treaty;

(e) agree upon procedures and dates for destruction or dismantling of ABM systems or their components in cases provided for by the provisions of this Treaty;

(f) consider, as appropriate, possible proposals for further increasing the viability of this Treaty; including proposals for amendments in accordance with the provisions of this Treaty;

(g) consider, as appropriate, proposals for further measures aimed at limiting strategic arms.

2. The Parties through consultation shall establish, and may amend as appropriate, Regulations for the Standing Consultative Commission governing procedures, composition and other relevant matters.

Article XIV

1. Each Party may propose amendments to this Treaty. Agreed amendments shall enter into force in accordance with the procedures governing the entry into force of this Treaty.

2. Five years after entry into force of this Treaty, and at five-year intervals thereafter, the Parties shall together conduct a review of this Treaty.

Article XV

1. This Treaty shall be of unlimited duration.

2. Each Party shall, in exercising its national sovereignty, have the right to withdraw from this Treaty if it decides that extraordinary events related to the subject matter of this Treaty have jeopardized its supreme interest. It shall give notice of its decision to the other Party six months prior to withdrawal from the Treaty. Such notice shall include a statement of the extraordinary events the notifying Party regards as having jeopardized its supreme interests.

Article XVI

1. This Treaty shall be subject to ratification in accordance with the constitutional procedures of each Party. The Treaty shall enter into force on the day of the exchange of instruments of ratification.

2. This Treaty shall be registered pursuant to Article 102 of the Charter of the United Nations.

Done at Moscow on May 26, 1972, in two copies, each in the English and Russian languages, both texts being equally authentic.

FOR THE UNITED STATES OF AMERICA:

RICHARD NIXON

President of the United States of America

FOR THE UNION OF SOVIET SOCIALIST REPUBLICS:

L. I. BREZHNEV

General Secretary of the Central Committee of the CPSU

Protocol to the Treaty between the United States of American and the Union of Soviet Socialist Republics on the Limitation of Anti-Ballistic Missile Systems

Signed at Moscow July 3, 1974
Ratification advised by U.S. Senate November 10, 1975
Ratified by U.S. President March 19, 1976
Instruments of ratification exchanged May 24, 1976
Proclaimed by U.S. President July 6, 1976
Entered into force May 24, 1976

The United States of America and the Union of Soviet Socialist Republics hereinafter referred to as the Parties,

Proceeding from the Basic Principles of Relations between the United States of America and The Union of Soviet Socialist Republics signed on May 29, 1972,

Desiring to further the objectives of the Treaty between the United States of America and the Union of Soviet Socialist Republics on the Limitation of Anti-Ballistic Missile Systems signed on May 26, 1972, hereinafter referred to as the Treaty,

Reaffirming their conviction that the adoption of further measures for the limitation of strategic arms would contribute to strengthening international peace and security,

Proceeding from the premise that further limitation of anti-ballistic missile systems will create more favorable conditions for the completion of work on a permanent agreement on more complete measures for the limitation of strategic offensive arms,

Have agreed as follow:

Article I

1. Each Party shall be limited at any one time to a single area out of the two provided in Article III of the Treaty for deployment of anti-ballistic missile (ABM) systems or their components and accordingly shall not exercise its right to deploy an ABM system or its components in the second of the two ABM system deployment areas permitted by Article III of the Treaty, except as an exchange of one permitted area for the other in accordance with Article II of the Protocol.

2. Accordingly, except as permitted by Article II of this protocol: the United States of America shall not deploy an ABM system or its components in the area centered on its capital, as permitted by Article III(a) of the Treaty, and the Soviet Union shall not deploy an ABM system or its components in the deployment area of intercontinental ballistic missile (ICBM) silo launchers as permitted by Article III(b) of the Treaty.

Article II

1. Each party shall have the right to dismantle or destroy its ABM system and the components thereof in the area where they are presently deployed and to deploy an ABM system or its components in the alternative area permitted by Article III of the Treaty, provided that prior to initiation of construction, notification is given in accord with the procedure agreed to in the Standing Consultative Commission, during the year beginning October 3, 1977 and ending October 2, 1978, or during any year which commences at five year intervals thereafter, those being the years for periodic review of the Treaty, as provided in Article XIV of the Treaty. This right may be exercised only once.

2. Accordingly, in the event of such notice, the United States would have the right to dismantle or destroy the ABM system and its components

in the deployment area of ICBM silo launchers and to deploy an ABM system or its components in an area centered on its capital, as permitted by Article III(a) of the Treaty, and the Soviet Union would have the right to dismantle or destroy the ABM system and its components in the area centered on its capital and to deploy an ABM system or its components in an area containing ICBM silo launchers, as permitted by Article III(b) of the Treaty.

3. Dismantling or destruction and deployment of ABM systems or their components and the notification thereof shall be carried out in accordance with Article VIII of the ABM Treaty and procedures agreed to in the Standing Consultative Commission.

Article III

The rights and obligations established by the Treaty remain in force and shall be complied with by the Parties except to the extent modified by this Protocol. In particular, the deployment of an ABM system or its components within the area selected shall remain limited by the levels and other requirements established by the Treaty.

Article IV

This Protocol shall be subject to ratification in accordance with the constitutional procedures of each Party. It shall enter into force on the day of the exchange of instruments of ratification and shall thereafter be considered an integral part of the Treaty.

DONE at Moscow on July 3, 1974, in duplicate, in the English and Russian languages, both texts being equally authentic.

FOR THE UNITED STATES OF AMERICA:

RICHARD NIXON
President of the United States of America

FOR THE UNION OF SOVIET SOCIALIST REPUBLICS:

L. I. BREZHNEV
General Secretary of the Central Committee of the CPSU

Agreed Statements, Common Understandings, and Unilateral Statements regarding the Treaty between the United States of America and the Union of Soviet Socialist Republics on the Limitation of Anti-Ballistic Missiles

I. Agreed Statements

The document set forth below was agreed upon and initialed by the Heads of the Delegations on May 26, 1972 (letter designations added);

AGREED STATEMENTS REGARDING THE TREATY BETWEEN THE UNITED STATES OF AMERICAN AND THE UNION OF SOVIET SOCIALIST REPUBLICS ON THE LIMITATION OF ANTI-BALLISTIC MISSILE SYSTEMS

[A]

The Parties understand that, in addition to the ABM radars which may be deployed in accordance with subparagraph (a) of Article III of the Treaty, those non-phased-array ABM radars operational on the date of signature of the Treaty within the ABM system deployment area for defense of the national capital may be retained.

[B]

The Parties understand that the potential (the product of mean emitted power in watts and antenna area in square meters) of the smaller of the two large phased-array ABM radars referred to in subparagraph (b) of Article III of the Treaty is considered for purposes of the Treaty to be three million.

[C]

The Parties understand that the center of the ABM system deployment area centered on the national capital and the center of the ABM system deployment area containing ICBM silo launchers for each Party shall be separated by no less than thirteen hundred kilometers.

[D]

In order to insure fulfillment of the obligation not to deploy ABM systems and their components except as provided in Article III of the Treaty, the Parties agree that in the event ABM systems based on other physical principles and including components capable of substituting for ABM interceptor missiles, ABM launchers, or ABM radars are created in the future, specific limitations on such systems and their components would be subject to discussion in accordance with Article XIII and agreement in accordance with Article XIV of the Treaty.

[E]

The Parties understand that Article V of the Treaty includes obligation not to develop, test or deploy ABM interceptor missiles for the delivery by each ABM interceptor missile of more than one independently guided warhead.

[F]

The Parties agree not to deploy phased-array radars having a potential (the product of mean emitted power in watts and antenna area in square meters) exceeding three million, except as provided for in Articles III, IV and VI of the Treaty, or except for the purposes of tracking objects in outer space or for use as national technical means of verification.

[G]

The Parties understand that Article IX of the Treaty includes the obligation of the U.S. and U.S.S.R. not to provide to other States technical description or blueprints specially worked out for construction of ABM systems and their components limited by the Treaty.

2. Common Understandings

Common understandings of the Parties on the following matters was reached during the negotiations:

A. Location of ICBM Defenses

The U.S. Delegation made the following statement on May 26, 1972:

Article III of the ABM Treaty provides for each side one ABM system deployment area centered on its national capital and one ABM system deployment area containing ICBM silo launchers. The two sides have registered agreement on the following statement: "The Parties understand that the center of the ABM system deployment area centered on the national capital and the center of the ABM system deployment area containing ICBM silo launchers for each Party shall be separated by no less than thirteen hundred kilometers." In this connection, the U.S. side notes that its ABM system deployment area for defense of ICBM silo launchers, located west of the Mississippi River, will be centered in the Grand Forks ICBM silo launcher deployment area. (See Agreed Statement [C].)

B. ABM Test Ranges

The U.S. Delegation made the following statement on April 26, 1972:

Article IV of the ABM Treaty provides that "the limitations provided for in Article III shall not apply to ABM systems or their components used for development or testing and located within current or additionally agreed test ranges." We believe it would be useful to ensure that there is no misunderstanding as to current ABM test ranges. It is our understanding that ABM test ranges encompass the area within which ABM components are located for test purposes. The current U.S. ABM test ranges are at White Sands, New Mexico, and at Kwajalein Atoll, and the current Soviet ABM test range is near Sary Shagan in Kazakhstan. We consider that non-phased array radars of types used for range safety or instrumentation purposes may be located outside of ABM test ranges. We interpret the reference in Article IV to "additionally agreed test ranges" to mean that ABM components will not be located at any other test ranges without prior agreement between our governments that there will be such additional ABM test ranges.

On May 5, 1972, the Soviet Delegation stated that there was a common understanding on what ABM test ranges were, that the use of the types of non-ABM radars for range safety or instrumentation was not limited under the Treaty, that the reference in Article IV to "additionally agreed" test ranges was sufficiently clear, and that national means permitted identifying current test ranges.

C. Mobile ABM Systems

On January 29, 1972, the U.S. Delegation made the following statement: Article V(1) of the Joint Draft Text of the ABM Treaty includes an undertaking not to develop, test, or deploy mobile land-based ABM systems and their components. On May 5, 1971, the U.S. side indicated that, in its view, a prohibition on deployment of mobile ABM systems and components would rule out the deployment of ABM launchers and radars which were not permanently fixed types. At that time, we asked for the Soviet view of this interpretation. Does the Soviet side agree with the U.S. side's interpretation put forward on May 5, 1971?
On April 13, 1972, the Soviet Delegation said there is a general common understanding on this matter.

D. Standing Consultative Commission

Ambassador Smith made the following statement on May 22, 1972:
The United States proposes that the sides agree that, with regard to initial implementation of the ABM Treaty's Article XIII on the Standing Consultative Commission (SCC) and of the consultation Articles to the Interim Agreement on offensive arms and the Accidents Agreement, agreement establishing the SCC will be worked out early in the follow-on Strategic Arms Limitation Talks (SALT) negotiations; until that is completed, the following arrangements will prevail: when SALT is in session, any consultation desired by either side under these Articles can be carried out by the two SALT Delegations; when SALT is not in session, ad hoc arrangements for any desired consultations under these Articles may be made through diplomatic channels.

Minister Semenov replied that, on an ad referendum basis, he could agree that the U.S. statement corresponded to the Soviet understanding.

E. Standstill

On May 6, 1972, Minister Semenov made the following statement:

In an effort to accommodate the wishes of the U.S. side, the Soviet Delegation is prepared to proceed on the basis that the two sides will in fact observe the obligations of both the Interim Agreement and Anti-ballistic Missile (ABM) Treaty beginning from the date of signature of these two documents.

In reply, the U.S. Delegation made the following statement on May 20, 1972:

The U.S. agrees in principle with the Soviet statement made on May 6 concerning observance of obligations beginning from date of signature, but we would like to make clear our understanding that this means that, pending ratification and acceptance, neither side would take any action prohibited by the agreements after they had entered into force. This understanding would continue to apply in the absence of notification by either signatory of its intention not to proceed with ratification or approval.

The Soviet Delegation indicated agreement with the U.S. statement.

3. Unilateral Statements

The following noteworthy unilateral statements were made during the negotiations by the United States Delegation:

A. Withdrawal from the ABM Treaty

On May 9, 1972, Ambassador Smith made the following statement:

The U.S. Delegation has stressed the importance the U.S. Government attaches to achieving agreement on more complete limitations on strategic offensive arms, following agreement on an ABM Treaty and on an Interim Agreement on certain measures with respect to the limitation of strategic offensive arms. The U.S. Delegation believes that an objective of the follow-on negotiations should be to constrain and reduce on a long-term basis threat to the survivability of our respective strategic retaliatory forces. The U.S.S.R. Delegation has also indicated that the objectives of SALT would remain unfulfilled without the achievement of an agreement providing for more complete limitations on strategic offensive arms. Both sides recognize that the initial agreements would be steps toward the achievement of more complete limitations on strategic arms. If an agreement providing for more complete strategic offensive arms limitations were not achieved within five years, U.S. supreme interests could be jeopardized. Should that occur, it would constitute a basis for withdrawal from the ABM Treaty. The U.S. does not wish to see such a situation occur, nor do we believe that the U.S.S.R. does. It is because we wish to prevent such a situation that we emphasize the importance the U.S. Government attaches to achievement of more complete limitations on strategic offensive arms. The U.S. Executive will inform the Congress, in connection with Congressional consideration of the ABM Treaty and the Interim Agreement, of this statement of the U.S. position.

B. Tested in ABM Mode

On April 7, 1972, the U.S. Delegation made the following statement:

Article II of the Joint Text Draft use the term "tested in an ABM mode," in defining ABM components, and Article VI includes certain obligation concerning such testing. We believe that the sides should have a common understanding of this phrase. First, we would note that the testing provisions of the ABM Treaty are intended to apply to testing which occurs after the date of signature of the Treaty, and not to any testing which may have occurred in the past. Next, we would amplify the remarks we have made on this subject during the previous Helsinki phase by setting forth the objectives which govern the U.S. view on the subject, namely, while prohibiting testing of non-ABM components for ABM purposes; not to prevent testing of non-ABM components for non-ABM purposes. To clarify our interpretation of "tested in an ABM mode," we note that we would consider a launcher, missile or radar to be "tested in an ABM mode" if, for example, any of the following events occur: (1) a launcher is used to launch an ABM interceptor missile, (2) an interceptor missile is flight tested against a target vehicle which has a flight trajectory with characteristics of a strategic ballistic missile flight trajectory, or is flight tested in conjunction with the test of an ABM interceptor missile or an ABM radar at the same test range, or is flight tested to an altitude inconsistent with interception of targets against which air defenses are deployed, (3) a radar makes measurements on a cooperative target vehicle of the kind referred to in item, (2) above during the reentry portion of its trajectory or makes measurements in conjunction with the test of an ABM interceptor missile or an ABM radar at the same test range. Radars used for purposes such as range safety or instrumentation would be exempt from application of these criteria.

C. No-Transfer Article of ABM Treaty

On April 18, 1972, the U.S. Delegation made the following statement:

In regard to this Article [ix], I have a brief and I believe self-explanatory statement to make. The U.S. side wishes to make clear that the provisions of this Article do not set a precedent for whatever provision may be considered for a Treaty on Limiting Strategic Offensive Arms. The question of transfer of strategic offensive arms is a far more complex issue, which may require a different solution.

D. No Increase in Defense of Early Warning Radars

On July 28, 1970, the U.S. Delegation made the following statement:

Since Hen House radars [Soviet ballistic missile early warning radars] can detect and track ballistic missile warheads at great distances, they have a significant ABM potential. Accordingly, the U.S. would regard any increase in the defenses of such radars by surface-to-air missiles as inconsistent with an agreement.

Author's Comments on Discussions Regarding the Treaty:

When President Reagan announced the SDI program on March 23, 1983, observers of arms control were concerned with its potential effect on the ABM Treaty. The recent push for "early deployment" of a partial SDI system has focused even more attention on the ABM Treaty and its interpretation. In order to understand the arguments of this current debate, it is important to review the context in which the ABM Treaty was negotiated.

The Importance of the ABM Treaty:

During the first Strategic Arms Limitation Talks (SALT I), both the United States and the Soviet Union recognized that neither side would agree to limits on nuclear weapons unless defensive systems were also restricted. Otherwise, nuclear weapons would be built up to counter any defenses and arms control would be impossible. The central purpose of the restrictions on defenses outlined in the ABM Treaty is clear, it was to permit SALT I to proceed with some chance of success. An "interim" agreement on limits for nuclear weapons reached in 1972 was to be followed by a more comprehensive treaty, SALT II, which was agreed to in 1977, but was never ratified by the U.S. Senate. As a result, there is currently no binding treaty that limits the deployment of additional nuclear weapons. Any restrictions on the United States or the Soviet Union are self-imposed for political, not legal, reasons. The ABM Treaty is the only legally binding arms control treaty in force today with no expiration date.

ABM Treaty Provisions:

In this context, the provisions of the ABM Treaty have been an important consideration in the subsequent development of strategic weapons. Article I (2) states, "Each Party undertakes not to deploy ABM systems for a defense of the territory of its country and not to provide a base for such a defense, and not to deploy ABM systems for defense of an individual region except as provided for in Article III of this Treaty. Article III as amended by the protocol agreement allows one site with "no more than one hundred ABM launchers and no more than one hundred ABM interceptor missiles."

The United States deployed Safeguard interceptor missiles with nuclear warheads at a site near Grand Forks, North Dakota. However, the economic costs outweighed the military benefits of a limited ABM system, so a decision was made to dismantle the Safeguard ABM system. The Soviet Union decided to maintain its one permitted ABM system around Moscow. Therefore, it is disingenuous to point out the Moscow ABM system as an argument in support of the SDI or as proof of Soviet belligerence. The Soviet system is in accordance with the Treaty.

SDI Plans Conflict With Both Intent and Wording of ABM Treaty:

Article I clearly prohibits deployment of comprehensive defenses originally envisioned by President Reagan when he announced the SDI program. Article III limits ABM deployment to 100 interceptor missiles at one site. However, since 1985, the Reagan Administration has been reinterpreting the ABM Treaty to justify development and testing of exotic technology such as lasers and particle beams. The treaty explicitly restricts development and testing, in addition to deployment, of ABM systems other than the single fixed land-based system permitted by Article III. The treaty specifically addresses the prospect of future ABM weapons based on technology other than "interceptor missiles."

Article V (1) states, "Each Party undertakes not to develop, test or deploy ABM systems or components which are sea-based, air-based, space-based, or mobile land-based." This provision clearly prohibits the "early deployment" of the partial defenses scenario for SDI recently embraced by Defense Secretary Caspar Weinberger and other members of the Reagan Administration."

Article II (1) states, "For the purpose of this Treaty an ABM system is a system to counter strategic ballistic missiles or their elements in flight trajectory, currently consisting of: (a) ABM missiles … (b) ABM launchers … and (c) ABM radars is "currently." In fact, it was the U.S. delegation that insisted on the word, and it was clear to both sides that it referred to the possibility of future technologies that both sides were researching."

At the treaty signing in 1972, the series of seven "Agreed Statements" which elaborated certain treaty provisions were initialed by the heads of the two delegations. The ABM treaty is a formal document signed by the heads of state. The "Agreed Statements" do not supplant or overrule provisions of the treaty. They were agreed to clarifications arrived at before the signing. The Reagan Administration has recently contended these statements supplant or override the Treaty in order to justify unlimited development and testing of exotic technologies such as lasers and particle beams.

Agreed Statement "D" states: "In order to insure fulfillment of the obligation not to deploy ABM systems and their components except as provided in Article III of the Treaty, the Parties agree that in the event ABM systems based on other physical principals and including components capable of substituting for ABM missiles, ABM launchers, or ABM radars are created in the future, specific limitations on such systems and their components would be subject to discussion in accordance with Article XIII and agreement in accordance with Article XIV of the Treaty.

The Reagan Administration is deliberately misinterpreting the ABM Treaty:

It is manifestly clear from the language of the ABM Treaty itself and also the supplemental Agreed Statement "D" that the Reagan Administration has no basis for its "reinterpretation." Agreed Statement "D" refers to deploy-ment as provided by Article III, which, as amended, allows only one fixed land-based ABM system. The U.S. instigated insertion of the word "currently" in describing ABM systems in Article II covers the "other physical principles" stated in Agreed Statement "D." Moreover, Article V explicitly prohibits development and testing, as well as deployment, of any ABM system that is not fixed land based. Therefore, while Agreed Statement "D" allows development and testing of exotic technologies for the fixed land-based system permitted by Article III, deployment is prohibited unless discussed by the Standing Consultative Commission (Article XII) and the ABM Treaty is amended (Article XIV). Thus development of land-based laser weapons are permitted but are not deployable except by agreement.

Behind the Push for "Early Deployment" of SDI:

The Reagan Administration's desired interpretation of the ABM Treaty would justify unlimited testing of exotic technology, including tests in space. However, deployment of spaced-based components of an ABM system would still be prohibited, especially if they included old technology-based interceptor missiles (Article V and Agreed Statement "D"). Therefore, in terms of the ABM Treaty, deployment is only part of the issue. Behind the push for "early deployment" of SDI systems is a desire by the Reagan Administration to conduct tests and deployments that would violate the ABM Treaty, thus destroy it. Attorney General Edwin Meese was perhaps the most candid when he said that the goal was to ensure that SDI would be "not tampered with by future administrations." By discrediting the ABM Treaty, the Administration would indeed limit the options of future administrations.

The political motivation of destroying the PALM Treaty has even damaged the SDI program itself. Funding to develop interceptor missiles has been diverted from research on exotic technologies. Requirements for the effectiveness of the interceptors has been lowered, in order to facilitate early deployment, to the point where even some SDI supporters are skeptical of the military utility of "early deployment." Adm. William Crowe, Jr., Chairman of the Joint Chiefs of Staff has said, "There are still many uncertainties involved with the weapon, and before any system can be deployed we must have a high degree of confidence these uncertainties are resolved." Finally, the plan outlined by Defense Secretary Caspar Weinberger for a limited deployment certainly would not fulfill President Reagan's vision of making nuclear weapons "impotent and obsolete." Although Weinberger counters that phased deployment is the only practical method, he has not specified how the pieces would fit together.

A3—COOPERATIVE VENTURES BETWEEN THE UNITED STATES AND THE SOVIET UNION

The exchanges between the United States and the Soviet Union may be described in three categories: business ventures, governmental exchanges and agreements and private person initiatives. The following is only a partial listing of efforts extending from the early 1970s.

1. BUSINESS VENTURES

In Nov. 1985 the Soviet Union established a Department for space applications called, "Glavkosmos of the U.S.S.R." Chairman of Glavkosmos is Alexander I. Dunayev who said in a statement through Novisti Press Agency that there is great interest in new space technology by the various ministries, departments and research organizations in the U.S.S.R. Dunayev said: "the potential of international cooperation in the development and peaceful uses of outer space is actually inexhaustible." (*Defense Daily*, November 21, 1985, p. 111.)

As reported in *Aviation Week and Space Technology*:

The Soviet Union's Glavkosmos has called on the U.S. to end restrictions that prevent most Western satellites from being launched on Soviet Vehicles.

Glavkosmos is focusing commercial launch service sales efforts, on multinational organizations such as Inmarsat and Intelsat, where it believes it has a better chance of success, according to Dunaev.

One of the Soviets' first commercial propositions was its 1983 offer to Inmarsat for the launch of the London-based organization's second-generation maritime communication space craft (AW&ST, June 20, 1983, p. 18).

— *Aviation Week and Space Technology*, "Soviets Seek Western Launch Bookings," October 20, 1986, p. 104

As reported in Defense Daily, "The Soviet Union told the United Nations last week that it is prepared to launch spacecraft of other nations and proposes establishment of a World Space Organization to oversee and promote joint international space projects while ensuring that space is used for peaceful purposes.

"A prerequisite for establishing the organization, which would be affiliated with the U.N., would be an agreement banning space weapons.

"In a letter to U.S. secretary General Javier Perez de Cuellar, Soviet Prime Minister Nikolai Ryzkhov said: 'The Soviet Union declares its readiness to exchange its accomplishments in outer space, to launch peaceful space vehicles of other countries and international organizations with Soviet carrier rockets on mutually acceptable terms.'

"The Soviets proposed that the World Space Organization be formed in conjunction with a world conference on space to be held by 1990. It would initially concentrate on communications, navigation, weather and remote sensing satellites, and then work to design and build space systems to investigate how to preserve the Earth's biosphere."

— *Defense Daily*, June 17, 1986, p. 263

As reported in The Washington Post, "The Soviet Union offered today to provide bargain space launches for Third World countries as part of a new profit-making service.

"'It is against our principles to cash in on others' errors and setbacks,' Premier Nikolai Ryzhkov said. 'Our proposal for launches of foreign spacecraft by Soviet carriers has been prompted by a desire to advance space exploration and use Soviet rockets and other space technology … efficiently.'"

— *The Washington Post*, "Soviets Offer Space Launches to Third World," January 6, 1987

As reported in the *The New York Times*, "A senior Soviet trade official today disclosed more details of the Soviet Union's plan to offer highly unusual operating independence and tax breaks as inducements to encourage American and other foreign companies to enter into joint business ventures with Soviet industries.

"Soviet trade officials said last month that 11 American companies had signed either formal or preliminary agreements to form joint ventures and 15 others were holding discussions.

"If the deals go through, the Monsanto Company will be a partner in a herbicide production plant in Alga, in the republic of Kazakhstan; the Occidental Petroleum Corporation will cooperate in secondary extraction from existing oilfields in the Volga region, and SSMC, Inc., formerly the Singer Sewing Machine Company, will join in the large scale production of sewing machines at a plant in the city of Orsha in the republic of Byelorussia."

> – Bill Keller, "Joint Ventures, Russian Style," *The New York Times,*
> January 6, 1987, p. D1, D6

"A change in Soviet regulations, scheduled to take effect in January, will allow many Soviet production organizations to talk trade with outsiders without going through the Ministry of Foreign Trade. This decentralization is expected to help small business people.

"The U.S. may ease some restrictions. Next May, the Department of Housing and Urban Development, which led the way in encouraging Soviet trade in construction goods in the 1970s, will co-sponsor a trade exhibition in the Soviet Union; 150 U.S. companies are expected to participate. And the Commerce Department says interagency discussions are under way about whether to substitute modified restrictions on the export of oil and gas equipment and technology when the current curbs expire Jan. 20."

> – *The New York Times*, "U.S. Small Businesses are Trying and Doing
> More Soviet Trading," Dec. 30, 1986, p. 1

The New York Times again: "The Soviets recognize the potential of space markets and have recently expanded their Glavkosmos commercial space organization to include marketing of three versions of their Proton launcher. In addition, Glavkosmos will also focus on marketing transponder relay capacity on their Gorizont communications satellites and exploring the market for Earth resources data collected by their remote sensing satellites. Soviet-developed or customer-supplied systems to perform materials processing and other micro-gravity experiments on Soviet space stations have also become available, according to Nicholas Johnson a Soviet space program expert with Teledyne Brown Engineering.

"Canadian and Soviet officials recently met to discuss both cooperative and commercial space ventures. Canada is now considering a Soviet offer to use the Proton rocket to launch commercial satellites. The two nations are also evaluating how to jointly develop and use remote sensing technology to monitor ice floes and map mineral resources. Advanced radar technology from the Soviet Cosmos 1500 series of satellites will probably be used in conjunction with developments with Canada's sophisticated radar program.

"In the United States, Art Dula, a Houston based lawyer has been marketing the Proton launcher. A number of innovative companies have also started to develop. The commercial External Tank Corp. has been formed in a joint project with a university consortium. External Tank Corp. is negotiating with the government to use discarded shuttle external tanks as low Earth orbit infrastructure. The tanks will be transformed into research modules, manufacturing facilities, or even an orbiting university. Pizza Hut recently concluded an agreement that will allow 100 of its stores to open in the Soviet Union beginning next year. Both Americans and Soviets will benefit from a new Soviet joint venture law which permits up to 49% Western ownership, several years of tax-free operations, complete independence from Soviet economic planning, and the freedom to experiment with Western labor-management. Details need to be worked out, but as Stephene Mazuerk, the head of the chemical company Monsanto's Moscow office, states, 'We have a start.'"

– Bill Keller, "Joint Ventures, Russian Style," *The New York Times*, January 6, 1987

"Already the Soviets have approximately 5,300 U.S. patents for their inventions, and the East bloc has earned about $80 million dollars in licensing fees and royalties from American businesses. Soviet electric welding and metallurgical processes have made significant inroads into

Western markets. Some of these processes are recognized as the most advanced in the world and have found use in the United States in everything from laying the subway rails that Pentagon officials ride to work to making nuclear fuel rods for the Department of Energy's plant at Hanford, Washington. John W. Kiser III, president of Kiser Research, contends that despite present barriers, "... the U.S.S.R. represents a vast, underutilized intellectual asset that we should be tapping. We tend to forget that Russians pioneered the science of chemistry and the development of lasers, put the first satellite and the first man in space ... Why shouldn't we take what we can from them."

> – Malcome Browne, "Technology From Russia Finds a Niche In
> U.S. Life," *The New York Times*, 16 December 1986

"'It's partly public relations and partly opportunism. But there is little doubt that the Soviet Union wants a bigger role in the world economy.' This is how economic analysts view a series of startling statements from Moscow in recent weeks. The Soviets have indicated interest in cooperating with the Organization of Petroleum of Exporting Countries (OPEC), the General Agreement on Tariffs and Trade (GATT), and the World Bank and International Monetary Fund (IMF).

"There are both short and long-term reasons for Moscow's interest. The main one, according to Soviet watchers: Moscow's borrowing needs have soared as oil prices have tumbled. Just to maintain the quantity of imports at last year's level, the Soviets will have to borrow $25 billion over the next five years, nearly doubling their gross hard currency debt, according to Jan Vanous, research director at Washington-based PlanEcon, Inc., which tracks Soviet-bloc economies. The money they need is in Western capital markets. The price for borrowing, Moscow may have calculated, will be more openness and cooperation with other nations."

> – *Christian Science Monitor*, 28 Aug. 1986, Reprinted in *Surviving Together; A Journal on Soviet–American Relations*, November 1986, Number 10, pgs. 74–75

"The U.S.–U.S.S.R. Agreement on Cooperation in the Field of Housing and other Construction, in separate but related development, the technical exchange program now has a commercial dimension. Secretary of Housing and Urban Development, Samuel R. Pierce, Jr., announced

June 10 that representatives of American firms in the construction and housing fields interested in doing business with the Soviet Union would meet in Washington, DC, on June 11 … approximately 200 U.S. businesses will attend."

– Surviving Together: An Update on Soviet–American Relations, July 1986, No. 9, p. 31–32

"Avon Products has signed an agreement with the Soviet Union to provide cosmetics for Soviet women in return for fine quality Soviet china and crystal, valued at $4 million."

– TASS, 27 Nov. 1985; Reprinted in *Surviving Together: An Update on Soviet–American Relations*, Feb. 1986, No. 8

2. GOVERNMENT EXCHANGE PROGRAMS

2.1 President Reagan's U.S.–Soviet Union Exchange Initiative

1. The final document for the exchange program was signed on November 21, 1985 at the Geneva Summit. At the same time President Reagan and General Secretary Gorbachev agreed in their Joint Statement to seek new ways to expand contacts between the two peoples.

2. As a result of this agreement a new office was created at the United States Information Agency, Coordinator for the President's U.S.–Soviet Exchange Initiative.

3. The exchange programs through the USIA can be broken down into three categories: long-term academic exchanges and the development of institutional relationships through grants for long and short-term projects; grants to professional groups for special projects; and a broad information program of television, radio, and printed contacts.

4. Benefits: To the extent that exchanges serve to open up, even a crack, the closed nature of the Soviet Union, they will help in breaking down stereotypes and misperceptions. Our own understanding of the Soviet Union has been greatly enhanced by exchanges, particularly long-term academic exchanges and repeat visits by American specialists in many fields.

Our professional cadre of Soviet specialists both inside and outside that government have testified in a variety of reports and studies to the importance of their exchange experiments in their professional development.

> – United States-Soviet Scientific Exchanges: Hearing before the Subcommittee on Europe and the Middle East, House Committee on Foreign Relations, Ninety-Ninth Congress, 2nd Session, 31 July 1986, pgs. 182–192

2.2 Other Government Exchanges

1. "U.S.-Soviet atomic energy officials sign protocol": U. S. and Soviet atomic energy officials signed a protocol marking revived formal exchanges on nuclear energy and adding nuclear plant safety to their discussions.

> – *Surviving Together: A Journal on Soviet–American Relations*, Nov. 1986, no. 10, p. 29, reprinted from *Journal of Commerce*, 25 August 1986

2. HUD and GOSTROY Promote Commercial Relations: As part of the U.S.–U.S.S.R. agreement on Cooperation in the Field of Housing and other construction, the U.S. Department of Housing and Urban Development (HUD), the lead U.S. agency in the agreement, has been working closely with representatives of the U.S.S.R. State Committee for Construction (GOSTROY), the U.S.S.R. Chamber of Commerce, and the U.S.–U.S.S.R. Trade and Economic Council to promote commercial relations between U.S. firms in the construction field and Soviet end-user ministries.

> – *Surviving Together; A Journal on Soviet American Relations,* Nov. 1986, No. 10, p. 30

3. USIA Arts America and the Soviet Embassy Report on Cultural Exchanges: A listing of performances in the U.S. and U.S.S.R.

> – *Surviving Together; A Journal on Soviet American Relations,* Nov. 1986, No. 10, p. 30

4. U.S. and Soviet Academy of Sciences Sign New Exchange Agreement: The U.S. Academy of Sciences (NAS) signed a new agreement on scientific cooperation with the Academy of Sciences of the U.S.S.R. in April 1986. The new agreement is limited to a two-year period and provides for a program of exchange of research scientists, a program of bilateral scientific workshops, exchange visits by Academy members, yearly meetings

of the officers of the two organizations, and development of cooperative research activities in non-sensitive fields.

    – *Surviving Together; A Journal on Soviet American Relations,* Nov. 1986, p. 31, No. 10

5. Tentative Agreement on U.S. Congress/Supreme Soviet Spacebridge: U.S. and U.S.S.R. policymakers have tentatively agreed to begin a series of 6–10 spacebridges on a variety of topics that would involve cooperative ventures, such as research in agriculture to feed the Third World, medical/scientific research, peaceful uses of space and environmental programs.

  * Soviet officials are expected to be in Washington this December to discuss the first such spacebridge, tentatively scheduled for early in 1987.

    – *Surviving Together; A Journal on Soviet American Relations,* Nov. 1986, p. 95, No. 10

2.3 U.S.–Soviet Scientific Exchanges: Government Agreements. Summary of findings by the Friends Committee on National Legislation;

2.3.1 General Environmental Issues

– Environmental Protection Agency:

Well over 2,000 Soviet and American scientists, engineers, and administrators have participated in exchange visits during the life of the Environmental Agreement. New data sets have been generated on natural aerosol properties in relatively pristine air on ecological baseline parameters in the Bering Sea, on the seismic activity induced by the filling of large reservoirs in mountainous regions, on the properties of trace gases at the ocean-atmosphere interface and on the distribution and migration of walrus and sea lion populations in the northeastern Pacific. Original joint research, or novel extensions of previous, have produced a wealth of published studies in both countries on a wide range of environmental topics: pollutant toxicity in fresh water, air pollution effects on forest vegetation, photochemical reactions of pesticides in natural waters, catalogs of rare and endangered plant species, reconstructions of prehistoric climate patterns, effects of solar activity on climate, and numerical studies of Earthquake source processes, to name a few.

– U.S. Fish and Wildlife Service:

Under the 1972 Environmental Exchange Program, U.S. Fish and Wildlife Service scientists met in Kharkov in October with Soviet scientists as part of annual working group discussions on environmental pollution and its effect on fish and wildlife. Discussions and symposia, ongoing since 1975, center on toxicological methodology and the behavioral characteristics of affected wildlife.

– National Academy of Sciences:

In 1976, an American scientist had first-hand exposure to remarkable Soviet successes in reversing the rapid decline of sturgeon populations. Dr. Ballard's glowing report describing the Soviet programs ultimately led the Wisconsin National Resources Department to test the Soviet methodology here in the U.S. in order to revive the dwindling sturgeon population in Lake Winnebago.

Dr. Viktor F. Petrenko, Senior Scientific Associate at the Institute for Solid State Physics in Moscow, spent three months in the U.S. under the auspices of the NAS–ASUSSR exchange program in 1985. Dr. Petrenko recently organized a 10-man research team on ice physics at the Institute of Solid-State Physics in Moscow, an effort which Dr. Offenbacher maintains is almost unparalleled in scale in the world scientific community. As a result of this NAS visit, the U.S. scientific community had access to a unique field in which the Soviets excel. Drs. Petrenko and Offenbacher submitted an abstract of a joint paper on the friction of ice which was recently accepted for presentation at an international meeting on the Physics and Chemistry of Ice in Grenoble.

– Department of Agriculture:

In the insect and disease area, a single visit in 1981 resulted in the collection of 1,500 gypsy moth larvae and parasites as part of research on natural enemies of this very devastating pest. At the time the agreement was suspended in 1982, very promising joint studies with the Soviets on bird and mammal predation of the gypsy moth had been conducted.

In the fire area, a highly maneuverable Soviet smokejumper parachute was obtained for testing under U.S. firefighting conditions. In harvesting, we exchanged techniques and designs for removal of small diameter trees. In the area of biogenocoenosis, basic research information and techniques for studying forest ecosystem dynamics was exchanged. In all areas, Soviet scientific literature which was previously unavailable or unknown to us was provided.

– Department of Transportation:

Activities included exchanges of information and delegations on air navigation research and development; i.e., air traffic production and use; airworthiness (certification of airplane designs); security, training and education; medical factors influencing crew performance; accident investigation procedures; and environmental factors.

– National Science Foundation:

Cooperative activities under the Electrometallurgy and Materials Working Group provided a number of clear scientific benefits to the United States. A large volume of materials was exchanged for testing and evaluation in the other country. The program provided an opportunity for U.S. scientists to observe advanced technologies in the U.S.S.R. in areas to which Soviet scientists had devoted more attention than have U.S. scientists, and to use and evaluate specialized equipment not available in the U.S. These include electroslag re-melting technology, which was developed by Soviet scientists, and plasma arc re-melting technology, where the U.S.S.R. had ten plasma arc furnaces and the U.S. none, thus enabling U.S. researchers to test the advantages of this technology without requiring the purchase of large, expensive melting equipment.

– Department of Housing and Urban Development:

While there are few general areas of Soviet technology with immediate application to the U.S., several special areas have been singled out by American private-sector participants as useful to them. These non-strategic technologies include construction in permafrost areas, Earthquake zone buildings, district-size and integrated utility systems, low energy use designs and technologies, and the use of large-scale models for research in building design.

2.3.2   Specifics and Benefits of U.S.–Soviet Cooperative Agreements

– Department of Agriculture:

Working Group on Forestry: May 24, 1972–June 1981 was the duration of the agreement.

1. Fire Development of effective methods and means of detection prevention, and control of forest fires.

2. Pests and Diseases—Integrated control of forest pests and diseases.

3. Biogeocoenosis—Classification of forest biogeocoenoses (ecosystems) and determination of their productivity.

4. Harvesting—Development of improved technological processes and means of mechanization of forest harvesting.

5. Reforestation—Reforestation and afforestation activities of shelterbelts; technology and large-scale mechanization of reforestation activities.

Benefits

"We consider that the Forest Service received benefits in all five areas covered by this exchange, and that these benefits easily justified the costs of participation."

June 19, 1973 Cooperative Agreement for the Department of Agriculture

1. Regular exchange of information, including forward estimates, on production, consumption, demand and trade of major agricultural commodities;

2. Methods of forecasting production, demand and trade of major agricultural products, including econometric methods;

3. Plant science, including genetics, breeding, plant protection and crop production, including production under semi-arid conditions;

4. Livestock and poultry science, including genetics, breeding, physiology, nutrition, disease protection and large-scale operations;

5. Soil science, including the theory of water movement, gases, salts, and heat in soils;

6. Mechanization of agriculture, including development and testing of new machinery, equipment and technology, as well as repair and technical service;

7. Application, storage and transportation of mineral fertilizers and other agricultural chemicals;

8. Processing, storage and preservation of agricultural commodities, including formula feed technology;

9. Land reclamation and reclamation engineering, including development of new equipment, designs and materials;

10. Use of mathematical methods and electronic computers in agriculture, including mathematical modeling of large-scale agricultural enterprises.

Benefits: Four Categories

1. Overall agricultural relations: The exchange program fosters contacts between U.S. and Soviet agricultural scientists, specialists and officials and creates a positive atmosphere for trade. The program has also increased the USDA cooperators' access to the Soviet Union.

2. Improved Economic Information: USDA and industry analysts have benefited from economic information gained through team visits to the Soviet Union. First-hand observation and discussions have helped answer basic production and utilization questions for important commodities. The publications exchange has assisted the National Agricultural Library in enlarging it: collection of published information about Soviet agriculture.

3. Scientific, Technical, and Institutional Knowledge: U. S. team visits to the Soviet Union have increased U.S. scientists and specialists' knowledge in several areas. The animal and plant protection and quarantine teams learned about the organization of these functions in a country as large and as the United States. Understanding Soviet quarantine policies and procedures is helpful to U.S. agricultural exporters who wish to ship products to the Soviet Union. The U.S. foresters who visited the Soviet Union were very interested in Soviet research, especially on seedling physiology, an area in which the U.S. is not currently doing extensive research. Information on Soviet research could strengthen the Forest Service research program and eventually improve reforestation programs in the United States.

4. Germplasm and Bio-Control Agents: The United States continually needs new germplasm for most of our important crops, because these crops are not native to our country. The Soviet Union, particularly in the areas near the Southern border, is potentially a prime source of new germplasm for U.S. agriculture, particularly for cereals, forages, and some fruits. Germplasm from cold-hearty tree varieties is also very important for windbreaks to control erosion and protect crops in the U.S. northern plain states. Likewise the Soviet Union is a potential source of bio-control agents for crop and forest pests such as gypsy moths.

– Department of Commerce, National Bureau of Standards:

1. The memorandum on cooperation was jointly signed by NBS Director Ernest Ambler and U.S.S.R. Academy Vice President Yevgeny Velikhov on December 13, 1978. The initial duration was for five years, and the period has since been extended for an additional five years, running now until December 12, 1988.

2. The memorandum of understanding initiated cooperative agreement in the following fields: Thermal physics and thermodynamics, materials science, spectroscopy, chemistry and chemical kinetics, and cryogenic science.

3. Benefits: Examination of these reports leads one to the conclusion that on the whole NBS has gained much information that would have not been available otherwise; that the U.S.S.R. scientists sent to the United States are technically competent and interested in cooperation … and that the exchange of information is roughly equivalent in the two directions.

– Department of Housing and Urban Development:

1. Signed on June 28, 1974 and renewed for five-year periods in 1979 and 1984.

2. There are six working groups involved in the agreement.

    a. Building design and Construction Technology

    b. Utilities systems

    c. Building materials and components

    d. Construction in Seismic areas

e. Construction under complicated climatic and geological conditions

f. Urban planning development and management

2. Benefits: As a result of its own internal review and survey of over sixty private sector participants, HUD, in late 1983, concluded that the Agreement should be renewed. This decision resulted from a weighing of benefits and costs of the agreement and an understanding of how it can be made more effective by focusing on more clearly defined technical areas.

– Department of Transportation:

1. Activities were formalized in June 1973, extended in 1978 for two years and three years in 1980. The agreement was allowed to expire in Sept. 1983 due to the downing of KAL 007.

2. Areas of Cooperation

a. Civil Aviation: Exchanges of information on air navigation R&D, general aviation production and use, airworthiness, security, training and education, medical factors influencing crew performance, accident investigation procedures, and environmental factors.

b. Marine transport: The program under this area included ice transiting; ocean commerce and cargo, ship equipment and crew training and measurement of sea pressure on ship hulls.

c. Bridge and Tunnel Construction

d. Rail: Exchanges included information on electrification, cold climate operations, freight management, track design maintenance, locomotive and other rail car designs, rolling stock and human factors.

e. Highway safety

f. Urban Transport

g. Transport of the Future: Exchanged information on magnetic levitation and high-speed rail technology.

h. Trade documentation

i. Hazardous Materials Transport

3. Benefits: Overall, Department officials believe the benefits gained under the lapsed Agreement were essentially worthwhile and well balanced.

– Environmental Protection Agency:

1. The agreement was reached on May 23, 1372, extended May 1977, and extended May 1982 for thirty-five years.

2. In its present form, the U.S.–U.S.S.R. agreement provides for cooperative activity in 38 specific projects among 11 general areas: air pollution, water pollution, pollution associated agriculture, the urban environment, nature conservation, marine pollution, biological/genetic effects, climatic effects Earthquake prediction, arctic/subarctic ecosystems, and legal administrative measures.

3. Benefits: On the most basic level, the Agreement is a unique channel of direct access to the growing environmental research and policy community in the U.S.S.R. … The knowledge and insights which this program affords to Soviet researchers is useful in substantiating the need for stricter environmental controls and a more comprehensive approach to environmental management in the U.S.S.R.

– National Aeronautics and Space Administration:

1. The Agreement ran from May 1972 through May 1982 with a new cooperative agreement pending.

2. Basic areas of cooperation

    a. Apollo–Soyuz Test Project

    b. Space biology and Medicine: Included the joint publishing of Foundations of Space Biology and Medicine, the standardization of key medical measurements and procedures, NASA flew experiments in three "Cosmos" unmanned dedicated biosatellite missions, and from 1980–1982, a joint program to measure and analyze bone calcium loss resulting from manned space flight.

    c. Near-Earth space, the Moon and Planets: Exchanges included information and data on lunar cartography, solar wind/planetary interaction, active magnetospheric experiments, Mars and Venus exploration, and * collisionless shock waves.

d. Study of Natural Environment: Efforts in this area focused on coordination surface, air, and space research over the oceans, and exchanging results of remote sensing measurements.

e. Space Meteorology: Bilateral cooperation took place in sounding rocket meteorology and satellite meteorology.

3. Benefits: There have been significant scientific benefits to the United States over the years from undertaken carefully constructed cooperative space projects with the Soviet Union. … Moreover, the access afforded to leading Soviet space scientists, engineers and research facilities gave the U.S. an understanding of Soviet capabilities.

* Types of activities included exchanges of scientists and technical information, joint development of research programs, joint research and testing and joint conference symposia.

– National Science Foundation:

1. The agreement lasted from 1972 to 1982.

2. General areas of cooperation were:

    a. Application of computers to management

    b. Chemical Catalysis

    c. Electrometallurgy and Materials

    d. Microbiology

    e. Physics

    f. Science policy

    g. Scientific and technical information

    h. Corrosion

    i. Heat and Mass Transfer

    j. Earth Sciences

    k. Polymer sciences

3. Benefits: An assessment conducted in 1984 by an external contractor concluded that the activities of four of the seven working groups established in the early '70s returned reasonable scientific benefits to the U.S. (Working groups 1,3,5,&6).

– U. S. Department of Interior:

1. The Cooperative agreement lasted from May 1972 to May 1982 with termination resulting from the Afghanistan invasion.

2. The 1972 S&T agreement was designed to meet four basic goals:

    a. Advancement of science and technology

    b. Enhancement of prestige of the U.S. and the U.S.S.R., reduction of tensions between the superpowers, and encouragement of international understanding.

    c. Promotion of trade between the two nations.

    d. Increased intergovernmental contact.

3. Program Activities included the following projects:

    a. Planning, Utilization and Management of Water Resources: The objective of this project involved comparative analysis of methods and equipment used to develop multipurpose utilization of water resources. Project aims also included development of a unified methodology for solving basic water economy problems while branches of the economy are under intense development.

    b. Program Two was Cold Weather Construction Techniques: This project proved to be the least successful and therefore no specific information has been published in the document.

    c. Program Three was Plastics in Hydrotechnical Construction:

    The objectives included development of new materials and techniques through comparison of technical levels and through cooperative research programs involving laboratory and field research.

    d. Program Four entailed Methods and Means of Automa-tion and Remote Control in Water Resource Systems: The principle objective of this project involved investigation of new control methods and algorithms for operating water resource systems.

4. Benefits: Include the potential for increased benefit to American industry, and continued cooperation between U.S. and Soviet team members for increased international understanding. With emphasis on the

joint study on improved design, construction, and operation of water resource projects, the joint study clearly has significant humanitarian benefits: improved food production and agricultural systems through better use of technology.

– Department of Health and Human Services:

1. The U.S.–U.S.S.R. Agreement on Cooperation in the Field of Medical Science and Public Health was signed May 23, 1972, extended for five years in 1977 and for five years in 1983 and will be extended for five years automatically unless announces its intention to terminate the agreement.

2. The Cooperative Agreement between the U.S. and U.S.S.R. on Artificial Heart Research and Development follows the same arrangement as above.

3. Specific areas of Cooperation:

    a. Cancer

    b. Cardiovascular Disease

    c. Artificial Heart

    d. Environmental Health (Coordinated with EPA)

    e. Arthritis

    f. Influenza, Acute Respiratory Diseases and Viral Hepatitis

    g. Mental Health

    h. Eye Diseases

    i. Biomedical Communications

    j. Individual Health Scientist Exchanges (on various topics)

4. Assessment of Benefits: Our general perception is that the costs of the exchanges and research have been, more or less, shared equally and the benefits to the health of both countries are also roughly comparable. (The report also notes) However, because no funds are specifically set aside for this cooperation, and instead activities must be justified and compete with other domestic and foreign research opportunities, we can say that nearly all the exchanges are undertaken with the clear expectation of benefit to the participating PHS components and to U.S. public health.

– Department of Energy:

1. The U.S.–U.S.S.R. Agreement on Cooperation in the Peaceful Uses of Atomic Energy was signed on June 21st, 1973, has renewed twice, and comes up for renewal in June 1988. The Agreement establishes a Joint Committee on Cooperation in the Peaceful Uses of Atomic Energy to review annually past cooperation, approve new proposals, and address important issues as they arise.

2. The Agreement listed three major areas for cooperation.

   a. Controlled Thermonuclear Fusion: The purpose of cooperation in controlled thermonuclear fusion is the eventual development of prototype and demonstration-scale thermonuclear reactors. Co-operation may include theoretical, calculational, experimental, and design-construction studies at all stages up the industrial-scale operation.

   b. Fast Breeder Reactors: Cooperation in fast breeder reactors is directed toward finding solutions to mutually agreed basic and applied problems connected with the design, development, construction and operation of nuclear power stations utilizing fast breeder reactors.

   c. Research in the Fundamental Properties of Matter: The purpose of cooperation in research on the Fundamental Properties of Matter is to include joint theoretical and experimental studies on mutually agreed subjects, and particularly in high, medium, and low energy physics, through utilization of accelerators, data processing equipment and other facilities of the two countries. Cooperation may also be undertaken on the design, planning and construction of joint facilities to be used in the area research.

3. Assessment of Benefits: In the fusion exchanges, DOE benefits substantially from access to the best of the U.S.S.R fusion scientists, experimental facilities and data complimentary to the U.S. program. DOE has gained considerably from the innovative ideas and approaches of the Soviets and has applied the Soviet experience in specific DOE experimental projects.

   In the area of FPM, The U.S. is a world leader and at present U.S. facilities are generally more advanced than those in the U.S.S.R. The Soviets seem to be positive on the exchanges. The trust and confidence in the exchanges seems to be on the upswing on the part of both the U.S. and Soviet participants. In FPM, the Soviets are continuing near the same level as in the past to send scientists to visit and work in the U.S. In fusion, the Soviets are now sending their best scientists to the U.S. and are allowing long-term exchanges to take place.

– Department of Commerce (NOAA):

1. The Agreement on Cooperation in Studies of the World Ocean (World Ocean Agreement) was signed in 1973. It was extended in 1978 and again in 1981, after it was suspended following the Soviet invasion of Afghanistan in 1979. In July 1985, it was extended for an additional three years, effective December 15, 1984.

2. Between 1974 and 1951 the National Marine Fisheries Service (NMFS) of NOAA was involved in a number of exchanges with the All-Union Scientific Institute of Marine Fisheries and Oceanography of the U.S.S.R. These exchanges did not take place under The World Ocean Agreement, and included joint symposia, workshops, visits, cruises, and working group meetings.

3. Under the Agreement, NOAA and Soviet scientists studied biological productivity and biochemistry. Biological studies, conferences, and scientist exchanges carried out under the agreement have contributed to our understanding of marine biological processes. Many activities under the Agreement involve other federal agencies. Two such projects were POLYMODE and international studies of the Southern Ocean. POLYMODE conducted by U.S. scientists under sponsorship of the National Science Foundation (NSF) and the Office of Naval Research (ONR), investigated the dynamics of large eddy circulation factors in the Northwest Atlantic. Soviet scientists also cooperated with the U.S. and other interested countries in the International Southern Ocean and possible impacts on other regions and climate. The Soviet Union also participated in the Deep Sea Drilling Projects, and may join with Canada, France, UK, FRG, Japan, European Science Foundation and the U.S. in the new Ocean Drilling Project.

4. Benefits: Both sides have benefited from cooperative activities under the Agreement. The U.S. delivered more in access to advanced scientific equipment and in project conceptualization and development; the

U.S.S.R. provided more ship time and some access for research purposes. The acquisition of data from cooperative ocean programs has been of great benefit to U.S. commercial, scientific, and defense interests.

– Scientific Cooperation Between the National Academy of Sciences and the Academy of Sciences in the U.S.S.R.:

1. The inter-academy science exchange program between the National Academy of Sciences and the Academy of Sciences of the U.S.S.R. dates from 1959 and has continued without interruption since that time … As a non-governmental program, it has enjoyed considerable stability, effectively bridging many low points in bilateral public relations. Finally, the program offers unique opportunities for individual U.S. scientists to develop cooperative activities tailored to their personal research interests.

2. The NAS program has the following objectives:

a. Advance U.S. science and technology interests by providing access to important geographic areas, institutions, individuals, and data sources in the U.S.S.R.;

b. Improve the basis for international scientific efforts and particularly efforts directed toward global problems by engaging key scientists from the U.S.S.R. in directly related bilateral activities;

c. Demonstrate new and improved modes of cooperation with the U.S.S.R. which can improve the effectiveness of science and technology interactions;

d. Improve U.S. understanding of U.S.S.R. science and technology capabilities, policies, and programs, and of the benefits and limitations of cooperation;

e. Enlarge the pool of U.S. scientists with expertise in science and technology activities in the U.S.S.R.;

f. Contribute to an improved atmosphere for U.S. political, economic, and cultural relations with the U.S.S.R.

3. Areas of Cooperation Involving Exchange Agreements

4. Benefits and Future Cooperation: In 1981, an NAS-hosted committee on U.S./U.S.S.R. scientific relations and inter-academy exchanges

chaired by Herbert York agreed that such exchanges are beneficial to the U.S. The Committee concluded that the formal scientific interactions between the NAS and the ASUSSR were justified on both scientific and political grounds, and that such interactions should certainly continue. In April 1986, the NAS signed a new agreement on scientific cooperation with the Academy of Sciences of the U.S.S.R.. The new agreement is limited to two years and provides for continuation of the exchange of individual scientists, a renewal of the program of bilateral scientific workshops, exchange visits by Academy members, yearly meetings of the officers of the two Academies, and development of cooperative research activities in non-sensitive fields. In addition, meetings of Academy officers will be held at least once a year to discuss problems and opportunities in fostering bilateral cooperation.

> – United States–Soviet Scientific Exchanges: A Hearing Before the Subcommittee on Europe and the Middle East, House Foreign Affairs Committee, 99th Congress, 2nd Session, 31 July 1986, pps. 94–225

3. PRIVATE INITIATIVES

Cultural exchanges are multiplying. In 1980, 12,922 visas were issued for visits to the Soviet Union. In 1985, the number of visas had risen to 45,049 and is still climbing. Space has made a significant contribution to breeding this form of cooperation and interchange. Satellite teleconferences have ranged from heart treatments to arms control issues. Even Phil Donahue broadcast his show for one week from the Soviet Union. Two U.S. companies provide Soviet television to American viewers. The Discovery channel plans to broadcast more than sixty hours of live Soviet television to watchers in the United States. The channel will take three-and-one-half hours of live programs a day from the Communications Satellite Co. (COMSAT). Already, U.S. and Soviet citizens communicate over the San Francisco-Moscow Teleport. The Teleport is a dedicated communications link that allows data and video to be transmitted between the United States and Soviet Union.

The Association of Space Explorers, an organization of astronauts and cosmonauts, consisting in 1986 of forty of the two hundred space flyers from sixteen of the eighteen space faring nations, began meeting in 1985 to exchange views and to promote space related educational projects. The Association maintains communications with its members in both the East and West by computer communications networks.

ONE-ON-ONE WITH DR. EDGAR MITCHELL

The following interview was conducted with Dr. Edgar Mitchell by the compiler, Carol Mersch, on October 15, 2005, at Mitchell's home in Lake Worth, FL during the George W. Bush administration.

Mersch: *When Foxes Guard the Hen House*, I have the manuscript here and it seems to be well-documented. It covers the Star Wars program that Reagan proposed. It speaks to how far you would go and how intensely you felt about it, because it took some balls to do that.

Mitchell: Because it seemed to exacerbate and move the world closer to a nuclear confrontation and nuclear exchange than it did in resolving the problem. And as I considered myself, correctly so I believe, an expert in the space environment and they were planning to put weapons in space—and a hair-trigger on a nuclear exchange—for the very simple reason that no nation in their right mind could allow a zero-delivery-time weapon, a laser weapon for example, orbiting over your facilities. If you really had hostile intent or were afraid, you had to shoot it down. And it was absolutely frightening to me that sane people would even *think* of such a thing.

Mersch: Was the intent further than just surveillance on Earth? Was it also to position ourselves for not only defense, but offence?

Mitchell: Remember we were still in the midst of the Cold War. The cracks in the communist system had not really become public knowledge and had not really emerged yet. So it was a confrontation between two superpowers. And a confrontation, in my opinion, that's too dangerous. It's just like what we call the "chicken hawks" of today—people who do not really understand the horror of war and don't really understand nuclear confrontation—poising two nations with their weapons pointed at each other, ready to fire and then taking it into space, so that you've reduced any margin for error whatsoever.

Mersch: Why do you think their purpose was and why were they doing that?

Mitchell: It's the mentality we're seeing right now. That power of control and war as a legitimate means of imposing your will on an opponent. But being a military officer and somewhat of an expert in nuclear weaponry, I realized it was pushing us right to the brink of brinkmanship and danger by taking weapons into space, taking the world right to the brink of destruction.

Mersch: Why did they do that at that particular time?

Mitchell: Because they are radicals just like the neo-cons we have now! They have ruined what we used to call Republican. The so-called neo-con or neo-conservative state isn't anything that anyone can recognize that used to be called Republican or Democrat. Those are long gone. Although I'm still a Republican, I'm embarrassed by what has happened to what's called Republicanism now, because its neo-con, it's neo-conservative, which is moving toward fascism. We're moving very rapidly toward a fascist state, and that's certainly not Republican or Democrat as any of us understand it. The very people that are managing the system are managing with a very erroneous, flawed mind-set as to where we should be going. They're taking us back to solving things with military force unilaterally as opposed to expanding multi-lateral, cultural openness. They're going exactly the wrong direction. So whereas we had the whole world with us at 9/11, this administration (George W. Bush) proceeded to blow all that in less than a year.

Mersch: Maybe they're going to suffer the same consequences. It's going to collapse under them. The king has no clothes.

Mitchell: Well, let's hope that's what happens before it's too late. We can't endure this another four years. It's going right toward a fascist regime. That's what they're doing. And if we don't get rid of Cheney, Wolfowitz, Rumsfeld, Condoleezza, Carl Rove, and a few others I could name, some of the project leaders . . .

Mersch: You didn't throw Bush W. in there.

Mitchell: He doesn't have much of a clue what's going on. He just signs whatever they ask him to. These guys allowed him to get elected and they are running the government. He doesn't have a clue what the hell is happening, I don't think. He couldn't do it when he was much younger, and he wasn't very bright then.

Mersch: Did you know him personally?

Mitchell: I knew him as part of the family. I knew him a little through George Sr. at their home at Hope Sound. I briefed George Sr. when he was Director of the CIA and first met him when he was Ambassador at the UN. I briefly met some of the children including George W. and wasn't impressed then—and am even less impressed now.

Mersch: What kind of briefing did you give to George Sr.?

Mitchell: It was when we had done the Uri Geller research at the Stanford Institute in the 1970s with the help of (Hal) Puthoff.[7] They were doing a CIA study at that time on remote viewing, so this was just another part of their project.

Mersch: Back when you were conducting the experiments? Why was he interested in that?

Mitchell: Right after the experiment. I think it was more to find out if I was really onto them, which I wasn't.

Mersch: What do you mean "on to them?"

---

7   Uri Geller/Hal Puthoff Stanford Research Studies on power of intention.

Mitchell: They had already funded and were supervising a remote viewing study at Stanford Research Institute (SRI), and they were setting up what we call Operation Stargate,[8] which was a remote viewing operation within CIA to observe the Soviet Union. I didn't know about the remote viewing project with CIA at that time. I subsequently learned about it, and I was asked to come back and brief George H.W. Bush, who was head of the CIA, on what we were doing. And I did it.

Mersch: They were feeling you out to find out what you knew about their own program?

Mitchell: I think that's true. But I didn't know. Of course the guys I was working with did know because they were running it. But they weren't leaking it to me.

Mersch: Was that all leading up to Reagan's whole deal with the Star Wars program?

Mitchell: Oh yes.

Mersch: At what point did you become aware that it was heading that direction?

Mitchell: What bothered me was, first, Jimmy Carter bringing fundamentalism into the White House. Then it became pretty obvious that Reagan was bringing in ultra right-wing politics. It didn't start to shape up to me what was happening until he announced the Star Wars program and Edward Teller[9] got into his ear—which was a very conservative think tank in Washington. They convinced Reagan to do Star Wars. I realized then that we were off on the wrong page. I guess what shocked me most, and what I wasn't prepared for, was how the economic system had been robbed and raped in the last 15 or 20 years—although I could see the handwriting on the wall with the huge CEO bonuses of hundreds of

---

8   Operation Stargate: a secret U.S. Army unit established by the Defense Intelligence Agency and SRI International to investigate the potential for psychic phenomena in military and domestic intelligence applications.
9   Edward Teller: theoretical physicist and advisor to Pres. Reagan on the "Star Wars" project; known as "father of the H-Bomb."

millions of dollars. I thought, God, this is obscene. I should have been able to see it. That's what Jeff Gates[10] clued me in on, how that had happened. The whole market system is coming apart.

Mersch: I'm reading where you say, "Braun said he believed these immense rockets would be capable of sending large payloads and men into space, but of course the U.S. military wasn't interested. What they wanted was the marriage of the two great military technologies that came out of WW2, the rocket engine, and the atomic bomb. They wanted ICBMs for the Cold War. They said exploration would have to wait." Was this the point in time that was beginning to happen? What made Reagan or anyone think there was a need for the Star Wars program?

Mitchell: Edward Teller was primarily the architect of that.

Mersch: Was it just a political thing?

Mitchell: Rabid anti-communist warped thinking. Edward Teller, who just died recently, and some of the ultra-conservative military strategists took the high ground—the high ground being space. It's an old military strategy going back forever, applying it to space—we needed to be in command of space and militarize it for purposes of weaponry and control.

Mersch: You must have been pretty steamed up about the time you wrote *When Foxes Guard the Hen House*.

Mitchell: Oh yes, I was pretty steamed up.

Mersch: How did you get hooked up with Carol Rosin?

Mitchell: She was Werner Von Braun's assistant when he left the government, and he was pretty disgusted. He went with Fairchild Industries. She was a schoolteacher and became his assistant. At some point he retired and left Fairchild before he died and told her to go see Edgar Mitchell. He's the only astronaut that knows what we're talking about. She was in a think tank in Washington for international cooperation in outer space and

---

10   Jeff Gates: member of the CIA and director of intelligence under President George H.W. Bush.

it was devoted to help us see—the Russians and ourselves—that competition in outer space was the wrong approach. So Carol Rosin was, and still is, an ardent proponent. And that's what we were both steamed about. She and her team were doing the research and I was doing the writing. And then when we presented the manuscript to publishers, 34 publishers, we couldn't get it published. Because the word got out and it got squelched.

Mersch: Was the material too sensitive right then?

Mitchell: Yes.

Mersch: And the manuscript was presented during the Reagan administration?

Mitchell: Yes.

Mersch: Then it was an exposé of sorts.

Mitchell: Oh yes.

Mersch: How did she get all the information and footnotes? There is some pretty specific stuff in the manuscript. She found documents where all those things were said and done?

Mitchell: Yes, she would dig it out.

Mersch: You mentioned that the CIA put a tap on your phone or put you under surveillance? What makes you think they did that?

Mitchell: I can't prove it. There was an agent or person in New York who was going to help me place it. I later became aware that they were on the other side. I was then placed under surveillance.

Mersch: How did you know that?

Mitchell: Their measures were not as undetectable in those days as they are now. My phone calls would have suspicious clicks on them or sounds that were indicative of a tapping operation. And then getting strange phone calls. Being questioned about things that only had taken place in a private discussion earlier with someone else. And recognizing it was going around in circles. And in the midst of all that I was being contacted by people who said they wanted to help me do it and then I suddenly realized it was a scam.

Mersch: It certainly was specific and appeared to be unarguable.

Mitchell: Well, the points I was making about polluting space, about the fact that it would put a hair-trigger on nuclear confrontation, all of that was true! But politicians didn't want to hear those things. The whole idea was the military precept to get there firstest with the mostest—but they don't take the broad philosophic view or look at what this is causing in terms of global dynamics. There were open letters to the President that up to 9,000 or 10,000 scientists signed onto that said don't do this!

Mersch: Did he know what he was doing anyway?

Mitchell: He was proceeding just like the idiots we're dealing with right now of a second-order conservative militarist mindset that the only way to control this is to go kill the bastards, and have more weapons, bigger weapons, faster weapons, commanding heights, than the other guy. That is the mentality. It is the utterly conservative radical military mentality— now and it always has been—from Napoleon on clear back to Alexander the Great. Get there the fastest with the mostest and wipe the bastards out—take what you want!

Mersch: Why did the program go away?

Mitchell: Because there was enough opposition. We made enough fuss that the research was done, but they could never make it operational. And that's what George W. is trying to do now—get back into the space business. It pretty much had to do with ground-based weapons against incoming war heads. The problem is you can't hit the goddamn things. They miss 'em. Well you can't hit 'em from space either. That's even worse. So the proper answer to the whole thing is nuclear disarmament. Really, the only answer is don't do it. But no one wants to hear that. It takes a totally different mindset to accept that. What I was trying to say with *Foxes* is that it escalates and almost assures the outcome you're trying to avoid.

Mersch: When I read it, just as the situation was beginning to flare up, it could have been written today. It has some analogies that were just incredible in it, with what is happening right now. It's almost a current book, except for who is saying what. Bush instead of Reagan. Is that correct?

Mitchell: Yes. The exact same mentality ratcheted it up a notch or two. And that is because the power of the Soviet Union has collapsed, and the United States is the sole dominant superpower. And so the leadership can say, "We can do what we goddamn well please and there's nobody to oppose us." And that ratchets it up.

Mersch: Isn't that the reason that the program went away too, though? Not only the opposition, but the fact that the Russian confrontation kind of itself just went away?

Mitchell: There was more opposition then. There were a lot of cross currents so that there was a lot of opposition then as opposed to what we have now where there's not even token opposition that's been thrown out. And the Soviets did not want the missile defense system. The whole notion of mutually assured destruction—we sit there and fire missiles at each other, and we're always afraid to fire them—is good enough. If you put them up there it just ratchets it up—another step closer toward a hair trigger problem. And the Soviets understood that. So working with them softened it a little.

Mersch: But I sense a feeling, in the U.S. anyway, with enough people are disgusted with Bush.

Mitchell: I wish it were true It listed all the news sources are listed, and sampling of people that get their news from those sources, and it showed that with Fox on one end up to PBS on the other, the percentages of misperceptions taking place in the public about Saddam being responsible for 9/11, being connected with Al Qaeda, etc. etc., and it starts with Fox Channel and the amount of false information believed by the public is up in the 80% to 90% from Fox and goes slowly down to . . .

Mersch: . . . to PBS?

Mitchell: That's exactly what happens. That explains why this administration has the popularity it does, because the people are simply misinformed because they prefer to listen to those news services.

Mersch: Is the CIA then a good thing or a bad thing?

Mitchell: When properly managed it's a good thing . . .

Mersch: You threw Rumsfeld in with the bunch.

Mitchell: Well, he's a radical! The problem is Cheney is even worse. They're both radical. . . . It seems that George Sr. thought Cheney would be a good Vice President for George W. because he (George W) knew so little about international politics, relations, etc. And Cheney simply turned out to be a turncoat. Everything that George Sr. wanted to happen, Cheney didn't want to happen. It's Cheney's ideas that are really running the government . . . he's a hard liner and managing Rumsfeld, Wolfowitz, that whole cabal.

Mersch: So, he's there with the puppet strings that are really running the show? You sure don't see his face very much.

Mitchell: Nope

Mersch: Well, you see George W. out there.

Mitchell: Well he's out there, but it's the (undistinguishable) behind running things.

Mersch: Well, you may see Cheney but not much.

Mitchell: He was the one over the CIA with Scooter Libby coloring all the reports.

Mersch: You mean the ones released by Colin Powell?

Mitchell: No, not by Colin Powell so much as the White House. That's where all this flap is coming from about false intelligence information ... whether it was distorted information. The spooks in CIA were generally pretty distraught that their stuff was being used and politicized—and used wrongly.

Mersch: It was being pumped through them and modified by Cheney? The speaker it came out of was Colin Powell. At least that's what everybody remembers.

Mitchell: Part of it was Colin Powell, but the contest between the State department and the Pentagon was severe, and Colin Powell put his reputation on the line and lost a lot of credibility ... because he was trying to be the good guy. But he ended up parroting the party line.

Mersch: Sure, but you mentioned they wanted you to brief him (George Sr.) primarily to determine if you knew anything about what they were doing.

Mitchell: Well, that was a later assessment by me. But I was invited to come and tell about the research we were doing in psychic research, remote viewing, psychokinesis, etc. But I think what they were really wanting to know was whether I was 'in' on what was being done in CIA by Puthoff while I was there, which I knew nothing about at the time.

Mersch: It would make me wonder why they would have such an extreme interest in the psychic end of things and remote viewing.

Mitchell: Well, it's not hard to understand, because it (remote viewing) was very strong in the Soviet Union, at least it was then. And our guys were checking it out in response to what was going on there. There was a book by Sheila Ostrander, *Psychic Discoveries Behind the Iron Curtain*,[11] that came out about that time.

Mersch: And you knew Russian, and you spoke Russian, right?

---

11 Psychic Discoveries Behind the Iron Curtain, Sheila Ostrander (Prentice Hall Trade, 1970, 1984) The existence and meaning of paranormal and parapsychological events in the 1970 investigation of Russian parapsychological research.

Mitchell: And I spoke Russian.

Mersch: Did you have an occasion to use it?

Mitchell: Well, a little bit when I was there, but I'd forgotten most of it by that time. I've been to the Soviet Union several times. I've been credited with—I don't think it's true—but I'm credited with getting Edward Naumov,[12] the parapsychologist, out of Soviet prison.

Mersch: Really. How did you do that?

Mitchell: I wrote letters on his behalf. He was subsequently assassinated in his own apartment.

Mersch: After he was released? Who else was working on that? Do you think you were instrumental?

Mitchell: I don't know. They credit me with it!

Mersch: Who? The Russians?

Mitchell: Yes, the Russians. I think my letters were passed on his behalf to the other Soviet scientists.

Mersch: Interesting. How did George Sr. come to know about your research?

Mitchell: I was doing the work with Uri Geller at SRI. They were doing a CIA project unbeknownst to me, so the information wasn't very clear, but what I was doing was no secret. I had sought sanction and help at political levels but never got any. So it was no secret what I was doing.

Mersch: I would think that NASA would be more interested, as would Washington—if they could peel off the bureaucratic layers and get back down to a meaningful level of what a space program would yield—if they

---

12   Edward Naumov. A Soviet parapsychologist sentenced to two years in a labor camp for accepting lecture fees without permission. "Émigré Tells of Research in soviet In Parapsychology for Military Use," *The New York Times*, Flora Lewis, June 19, 1977.

would draw on the very limited resource of you guys who devoted pretty much your entire life to it.

Mitchell: Well, yes and no. Some yes and some no. That really depends on whether you're talking about technicians or philosophers. Most of the guys are pretty much technicians.

Mersch: Well it takes a bit of both. You have to understand the technicalities of it, but you need to contribute in a philosophic manner. I know you're a philosopher, Edgar. It would appear to me, though, that using you guys on an advisory board would help them make better decisions and stay properly focused. If you could do that, and if anyone in Washington or NASA would listen to you on that, what would be your suggestion?

Mitchell: It's trying to find a balance between the unmanned probes and then eventually sending humans. But most of the scientific community is not too hot on sending humans because all they're interested in is data. And very few of them are interested in human experience. And you do have to put all of this in the context of the evolution of life on Earth. We're only doing that now as a result of the space program, and the Hubble telescope, and astronomy . . . starting to get a decent handle on what type of universe we're in. We really didn't know that, and we still don't know that. But the concentration on the space program has helped us to get an understanding of that. And most people aren't interested. Most people really don't care. They've got their pet ideas, and they've got their religious ideas, and what they've learned in history books when they were going through school, and that's all they're interested in. There are not very many real scholars in the general public burning to know these deep questions. But science and evolution is vastly dependent on that, and in due course we will go into space. We will go into deep space. We will go outside of our solar system. We will at least cursorily explore the other planets. But that's a hundred, or two hundred, or five hundred, or a thousand years away. I think with concerted effort we could be ready to go outside our solar system by the end of this century, but if we goof around and don't—we'll never do it. It's a big step to go outside of our solar system and we don't have the science to do it yet—and certainly not the technology. So there's a lot of new inventions and new creations and new science that has to be developed, and some people are working on that. But the politics of science is very restrictive and has as much political chaos as other forms of politics. It's inhibiting their progression. It's my personal opinion that we will never really solve the scientific issue of an ultimate theory of things until we have a theory of consciousness. That's missing. The scientific community is only in our time starting to address that properly. And still really haven't started to address it.

Mersch: But if there is a focus to do that, just like there was a focus to get to the Moon, and some kind of commitment from the government to do that—where there's focus to do that, you could certainly expedite the technology.

Mitchell: Yes, it would. But mostly our focus is driven by the corporate world—the economic model—and that's in producing money . . . and producing money for fewer and fewer people, not bolstering the sustenance and affluence of the world. So in my opinion the model is heading us in the wrong direction. It's going to run its course until it falls apart or something changes it.

Mersch: Well technology certainly seems to be focused on improving weaponry and what it takes to protect our little world.

Mitchell. Through the means of military force, yes, when there's really better ways of doing it than that. But no one's focused on that.

Mersch: Better ways to do it than what?

Mitchell: Than violence. Better than violence and military action.

Mersch: We're pretty base—pretty elementary.

Mitchell: We would say it's elementary, but not everyone agrees with that.

Mersch: I know they don't. But hand-to-hand combat?! People driving around in little tanks trying to shoot each other!?

Mitchell: If you're going to do that, you might as well use bows and arrows and knives.

—||—

ABOUT THE AUTHOR
NAVY CAPTAIN DR. EDGAR MITCHELL

Edgar Dean Mitchell was a United States Navy officer and aviator, test pilot, aeronautical engineer, and NASA astronaut. As the Lunar module pilot of Apollo 14 in 1971, he spent nine hours working on the lunar surface in the Fra Mauro Highlands region and was the sixth person to walk on the Moon.

A brilliant astronaut with a reported IQ of 180, Mitchell earned his Bachelor of science degree in industrial management from Carnegie Institute of Technology and entered the United States Navy in 1952. In 1961, he received his second bachelor's degree, in aeronautical engineering, from the U.S. Naval Postgraduate School.

Three years later, he earned his doctorate in Aeronautics and Astronautics from the Massachusetts Institute of Technology (MIT) where his thesis, "Guidance of Low-Thrust Interplanetary Vehicles," posited an elegant engineering theory for launching and flying to another planetary body with such a precise trajectory that it would consume a minimum amount of fuel. The theory would map the trajectory from the initial point to the final destination, the Moon, Mars, and beyond.

In 1966, Mitchell was selected as part of NASA's fifth astronaut group, where he served as lunar module pilot on Apollo 14, landing on the Moon with Alan Shepard aboard the lunar module *Antares* on February 5, 1971.

As a computer-literate astronaut, Mitchell was well suited for the missions of the Apollo spacecraft, whereas many of the other astronauts were not. Most of his fellow astronauts were engineers. For engineers, the journey to the Moon was an engineering problem to be solved. For scientists such as Mitchell, the journey to the Moon was the start of understanding the solar system. His global perspective of space, sustainability, and consciousness remained with him for the rest of his career.

On March 1, 1971, Mitchell was awarded the Presidential Medal of Freedom by President Richard M. Nixon. In 1973, he founded the Institute of Noetic Sciences (IONS), dedicated to the research of global awareness and universal consciousness. The program has since grown to over 30,000 members from 74 countries worldwide.

In 1985, Mitchell was instrumental in founding the Association of Space Explorers, an organization of astronauts from around the world who have flown in space, formed "to encourage international cooperation in the exploration of science and space and to foster greater environmental awareness."

On October 4, 1997, Mitchell was inducted into the Astronaut Hall of Fame. In 2005, he was nominated for the Nobel Peace Prize for his contribution to the study of universal awareness and global sustainability. In 2011, he received the Leonardo DaVinci Award for the Study of Thinking, presented to individuals who have made outstanding contributions to the history of technology through research, teaching, publications, and service to society.

After retiring from NASA in 1972, Mitchell became involved in scientific studies that brought him in contact with ranking members of government, CIA, FBI, and the Pentagon. Of specific interest to Mitchell was the suspected government cover up of the alleged UFO crash in his hometown near Roswell, New Mexico. A later death-bed confirmation by a witness convinced him the 1947 UFO incident was genuine. His suspicions were confirmed when he obtained a July 8,1947, FBI teletype stating the object found resembled a high-altitude weather balloon with a radar reflector, "but that telephonic conversation between their [headquarters] office and Wright field had not borne out this belief."

Mitchell publicly expressed his opinions that he was "90 percent sure" that many of the thousands of unidentified flying objects recorded since the 1940s belong to visitors from other planets.

"We all know that UFOs are real," he said, "now the question is where they come from."

Dr. Mitchell passed away February 4, 2016.

ABOUT VON BRAUN'S PROTÉGÉ
DR. CAROL ROSIN

Carol Rosin is a nationally recognized missile defense consultant and has consulted with a number of companies, organizations, government departments and the intelligence community.

She has testified before the U.S. Congress, the U.S. Senate, the President's Commission on Space, and has met with people in over 100 countries about the feasibility of banning space-based weapons. She received Bachelor's of Science from the University of Delaware and an honorary doctorate from Archbishop Solomon Gbadebo of the Orthodox College in Nigeria.

She is executive director of the Peace and Emergency Action Coalition for Earth, PEACE Inc., and the IDEA Foundation, as well as a world peace ambassador for the International Association of Educators for World Peace.

She is the first woman to hold an executive position in the aerospace industry at Fairchild Industries and was a close associate of the late Wernher von Braun, the father of modern rocketry, in the years before his death. In her time at Fairchild, Rosin served as the spokesperson for von Braun, with whom she created the film and educational program "It's Your Turn" to expand the diversity of people working in science fields. The program won many awards, including the Aviation Writers Award and the Science Teachers Gold Medal.

Rosin led an important political and public relations campaign to stop Ronald Reagan's Strategic Defense Initiative (SDI), his "Star Wars" proposal for the placement of laser weapons in space to defend the U.S. from Russians, identified terrorist nations, and von Braun's denoted "last enemy," invading aliens.

She was a world peace ambassador and was instrumental in creating satellite educational programs in the U.S. that offered the first two-way audio and visual training programs for more than 20 countries. She worked with John Podesta, former Clinton White House Chief of Staff, and counselor to President's Biden and Obama, to create the Treaty on the Prevention of Placement of Weapons in Outer Space.

Rosin helped set up the ATS-6 satellite educational and medical training programs across the U.S. in Appalachia, the Rocky Mountains, Alaska, and India, including the first national and international two-way, audio-visual satellite educational programs, including 5,000 villages in India.

She speaks regularly to various groups, including the CIA and U.S. Congress, warning that weapons in space would threaten our planet's delicate balance.

In 1977, after von Braun's death, she was hired as a space and missile defense consultant by such companies as TRW, where she worked on the MX missile, the space shuttle, and weather satellites. While working at TRW, she learned that the aerospace defense industries were preparing to escalate the arms race into space. The plan was to put hundreds of battle stations in space "with thousands of weapons pointed down our throats," a program diametrically opposed to von Braun's objectives.

During this time, Rosin became the National Chancellor of the International Association of Educators for World Peace (IAEWP) in consultative status with the UN Economic and Social Council. After bringing several experts to the UN Second Special Session of the Peaceful

Uses and Exploration of Outer Space in Vienna, Austria, Rosin founded the Institute for Security and Cooperation in Outer Space (ISCOS) in 1983.

She continues her vigorous campaign to educate political and administrative officials, aerospace industries, and the public on the potential repercussions in U.S. deployment of space-based weapons and the pressing need for international treaties to avert such outcomes.

"I do not oppose the militarization of outer space," she said. "I oppose the weaponization of outer space, and think that we should take another look to see if there are technologies other than the dangerous nuclear technologies now being used and planned for in space."

ABOUT THE COMPILER
CAROL MERSCH

Carol Mersch is an Oklahoma author and journalist specializing in narrative nonfiction. She has published and co-published over ten books and numerous media articles in the areas of space exploration, law enforcement, and spirituality.

Her close friendship with Apollo 14 astronaut Edgar Mitchell led her to develop *The Apostles of Apollo: The Journey of the Bibles to the Moon* (Pen-L Publishing, 2010), for which she was accepted into the Mayborn Literary Guild, and *The Space Less Traveled* (Pen-L Publishing, 2013), a book of quotations gleaned from her years of companionship with Mitchell, and *We Are One* (Pen-L Publishing, 2020), Mitchell's personal account of his transformative experience on his return to Earth onboard Apollo 14. In 2013, her literary document "Religion, Space Exploration and Secular Society" was accepted by Taylor & Frances, a national consortium in the UK offering document subscription services used by museums, libraries, and universities, including the Smithsonian Air & Space Museum.

Other recent books include *Undaunted: The Unflinching Faith, Audacity, and Ultimate Betrayal of Reverend John Maxwell Stout* (Pen-L Publishing, 2018), covering the life of the legendary NASA Chaplain responsible for landing of the first Bible on the Moon, and *Guilty When Black: One girl's journey down the twisted road of injustice & the atrocities of female incarceration* (Yorkshire Publishing, 2020), which follows the egregious imprisonment of an innocent young black girl at the hands of Oklahoma's unforgiving justice system.

Before launching her writing career, she served at the executive level of two Fortune 1000 enterprises and at the helm of three privately held companies where she received local and national recognition for her contributions to community and civic endeavors.

Mersch has authored numerous articles for national trade and online publications on information technology and developed an IT strategy manual utilized by several leading corporations and local governments. In 1999, her firm, Mersch-Bacher Associates, was awarded the Blue Star Award for entrepreneurship. The success of Mersch and her company was featured on a nationally televised PBS special. In August 2000, *The New York Times* cited her work in community and civic endeavors.

In 2004, she left the corporate world to form ProvidenceWorks LLC, a business enterprise for developing articles and books "that make a difference."

She has been featured on Houston Fox26, Tulsa NewsOn6, BBC World Radio, Dallas CBS Radio KRLD, MSNBC, and CNN Faith and also in two magazines in Europe, *Spaceflight Magazine* and *Sorted*, a Christian men's magazine, for her research into the first lunar Bible covered in her book *The Apostles of Apollo.* The historic Bibles were carried to the Moon and their heirship have been featured by the Associated Press, the *Houston Chronicle*, the *Baytown Sun,* MSNBC, Fox News, CNN Belief, and *Al Jazeera*'s "America Tonight."

In 2023 she was selected for the 76th edition of *Who's Who in America* for her contributions to civic and national endeavors. She was previously listed in *Who's Who in Science and Engineering* and *Who's Who in Finance and Engineering.*

PHOTO GALLERY

Wernher von Braun, German-American NASA rocket scientist who developed the Saturn V super heavy-lift launch vehicle that carried Apollo spacecraft to the Moon. Courtesy: NASA

Carol Rosin and Wernher von Braun at Fairfield Industries, 1974–1977. Source: Dr. Steven Greer, "Disclosure Project," National Press Club, Washington DC, 2001.

President Ronald Reagan and Hungarian theoretical physicist Edward Teller, "father of the hydrogen bomb," when Teller was awarded the National Medal of Science in 1983. Source: White House Photographic Collection.

Apollo 14 Navy Capt. Dr. Edgar Mitchell, sixth man to walk on the Moon on February 5, 1971. Courtesy: NASA

Apollo 14 Stuart Roosa, Alan Shepherd, and Edgar Mitchell after splashdown
in the south Pacific Ocean, February 9, 1971. Courtesy: NASA.

President John F. Kennedy and Dr. Wernher von Braun at NASA's Marshall
Space Flight Center in 1962. Courtesy: NASA

Walt Disney and Wernher von Braun, seen in 1954, collaborated on a
series of three educational films. Courtesy: NASA

Roswell
(1 page)

TELETYPE

FBI DALLAS   7-8-47   6-17 PM

DIRECTOR AND SAC, CINCINNATI   URGENT

FLYING DISC, INFORMATION CONCERNING.   HEADQUARTERS

EIGHTH AIR FORCE, TELEPHONICALLY ADVISED THIS OFFICE THAT AN OBJECT

PURPORTING TO BE A FLYING DISC WAS RE COVERED NEAR ROSWELL, NEW

MEXICO, THIS DATE. THE DISC IS HEXAGONAL IN SHAPE AND WAS SUSPENDED

FROM A BALLON BY CABLE, WHICH BALLON WAS APPROXIMATELY TWENTY

FEET IN DIAMETER.   FURTHER ADVISED THAT THE OBJECT

FOUND RESEMBLES A HIGH ALTITUDE WEATHER BALLOON WITH A RADAR

REFLECTOR, BUT THAT TELEPHONIC CONVERSATION BETWEEN THEIR OFFICE

AND WRIGHT FIELD HAD NOT   BORNE OUT THIS BELIEF. DISC AND

BALLOON BEING TRANSPORTED TO WRIGHT FIELD BY SPECIAL PLANE FOR EXAMINATI

INFORMATION PROVIDED THIS OFFICE BECAUSE OF NATIONAL INTEREST IN CASE .

AND FACT THAT NATIONAL BROADCASTING COMPANY, ASSOCIATED PRESS, AND

OTHERS ATTEMPTING TO BREAK STORY OF LOCATION OF DISC TODAY.

ADVISED WOULD REQUEST WRIGHT FIELD TO ADVISE CINCINNATI

OFFICE RESULTS OF EXAMINATION. NO FURTHER INVESTIGATION BEING

CONDUCTED.

WYLY
RECORDED
END

CXXXX ACK IN ORDER   EX-29   23 JUL 22 1947

UA 9 FBI CI MJW

DPI H8

8-38 PM O

6-22 PM OK FBI WASH DC

OK FBI OK

FBI teletype July 8,1947 reporting conflicting accounts of UFO debris found in Roswell, NM. Source: Dr. Edgar Mitchell

Dear Reader,
If you enjoyed this book enough to review it for Goodreads, B&N, or Amazon.com, I'd appreciate it!
Thanks, Carol

Find more great reads at
Pen-L.com

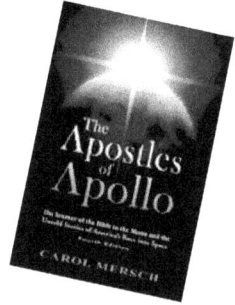

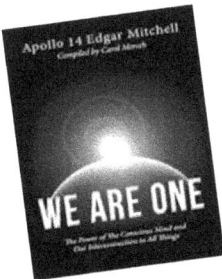

319

NOTES